# CALCULUS
# SUCCESS
## in 20 Minutes a Day

# CALCULUS SUCCESS
## in 20 Minutes a Day

Christopher Thomas

LEARNINGEXPRESS®

NEW YORK

Library of Congress Cataloging-in-Publication Data:
Thomas, Christopher, 1973-
    Calculus success in 20 minutes a day / Christopher Thomas.
        p. cm.
      ISBN 1-57685-536-8
    1. Calculus—Problems, exercises, etc. I. Title.
QA303.2.T47 2005
515—dc22                                                          2005027519


Printed in the United States of America

9 8 7 6 5 4 3 2 1

ISBN 1-57685-536-8

For information on LearningExpress, other LearningExpress products, or bulk sales, please write to us at:
    LearningExpress
    55 Broadway
    8th Floor
    New York, NY 10006

Or visit us at:
    www.learnatest.com

# About the Author ▶

**Christopher Thomas** is a professor of mathematics at the Massachusetts College of Liberal Arts. He has taught at Tufts University as a graduate student, Texas A&M University as a postdoctorate professor, and the Senior Secondary School of Mozano, Ghana, as a Peace Corps volunteer. His classroom assistant is a small teddy bear named $e^x$.

# Contents

# Introduction ▶

If you have never taken a calculus course, and now find that you need to know calculus—this is the book for you. If you have already taken a calculus course, but felt like you never understood what the teacher was trying to tell you—this book can teach you what you need to know. If it has been a while since you have taken a calculus course, and you need to refresh your skills—this book will review the basics and reteach you the skills you may have forgotten. Whatever your reason for needing to know calculus, *Calculus Success in 20 Minutes a Day* will teach you what you need to know.

## ▶ Overcoming Math Anxiety

Do you like math or do you find math an unpleasant experience? It is human nature for people to like what they are good at. Generally, people who dislike math have not had much success with math.

If you have struggles with math, ask yourself why. Was it because the class went too fast? Did you have a chance to fully understand a concept before you went on to a new one? One of the comments students frequently make is, "I was just starting to understand, and then the teacher went on to something new." That is why *Calculus Success* is self-paced. You work at your own pace. You go on to a new concept only when you are ready.

When you study the lessons in this book, the only person you have to answer to is you. You don't have to pretend you know something when you don't truly understand. You get to take the time you need to understand everything before you go on to the next lesson. You have truly learned something only when you thoroughly understand it. Take as much time as you need to understand examples. Check your work with the answers and if you don't feel confident that you fully understand the lesson, do it again. You might think you don't want to take the time to go back over something again; however, making sure you understand a lesson completely may save you time in the future lessons. Rework problems you missed to make sure you don't make the same mistakes again.

## ▶ How to Use This Book

*Calculus Success* teaches basic calculus concepts in 20 self-paced lessons. The book includes a pretest, a posttest, 20 lessons, each covering a new topic, and a glossary. Before you begin Lesson 1, take the pretest. The pretest will assess your current calculus abilities. You'll find the answer key at the end of the book. Each answer includes the lesson number that the problem is testing. This will be helpful in determining your strengths and weaknesses. After taking the pretest, move on to Lesson 1, *Functions*.

Each lesson offers detailed explanations of a new concept. There are numerous examples with step-by-step solutions. As you proceed through a lesson, you will find tips and shortcuts that will help you learn a concept. Each new concept is followed by a practice set of problems. The answers to the practice problems are in an answer key located at the end of the book.

When you have completed all 20 lessons, take the posttest. The posttest has the same format as the pretest, but the questions are different. Compare the results of the posttest with the results of the pretest you took before you began Lesson 1. What are your strengths? Do you have weak areas? Do you need to spend more time on some concepts, or are you ready to go to the next level?

## ▶ Make a Commitment

Success does not come without effort. If you truly want to be successful, make a commitment to spend the time you need to improve your calculus skills.

So sharpen your pencil and get ready to begin the pretest!

# CALCULUS
# SUCCESS
## in 20 Minutes a Day

# Pretest

Before you begin Lesson 1, you may want to get an idea of what you know and what you need to learn. The pretest will answer some of these questions for you. The pretest is 50 multiple-choice questions covering the topics in this book. While 50 questions can't cover every concept, skill, or shortcut taught in this book, your performance on the pretest will give you a good indication of your strengths and weaknesses.

If you score high on the pretest, you have a good foundation and should be able to work your way through the book quickly. If you score low on the pretest, don't despair. This book will take you through the calculus concepts, step by step. If you get a low score, you may need to take more than 20 minutes a day to work through a lesson. However, this is a self-paced program, so you can spend as much time on a lesson as you need. You decide when you fully comprehend the lesson and are ready to go on to the next one.

Take as much time as you need to do the pretest. When you are finished, check your answers with the answer key at the end of the pretest. Along with each answer is a number that tells you which lesson of this book teaches you about the calculus skills needed for that question. You will find the level of difficulty increases as you work your way through the pretest.

## ANSWER SHEET

| | | | |
|---|---|---|---|
| 1. | ⓐ | ⓑ | ⓒ | ⓓ |
| 2. | ⓐ | ⓑ | ⓒ | ⓓ |
| 3. | ⓐ | ⓑ | ⓒ | ⓓ |
| 4. | ⓐ | ⓑ | ⓒ | ⓓ |
| 5. | ⓐ | ⓑ | ⓒ | ⓓ |
| 6. | ⓐ | ⓑ | ⓒ | ⓓ |
| 7. | ⓐ | ⓑ | ⓒ | ⓓ |
| 8. | ⓐ | ⓑ | ⓒ | ⓓ |
| 9. | ⓐ | ⓑ | ⓒ | ⓓ |
| 10. | ⓐ | ⓑ | ⓒ | ⓓ |
| 11. | ⓐ | ⓑ | ⓒ | ⓓ |
| 12. | ⓐ | ⓑ | ⓒ | ⓓ |
| 13. | ⓐ | ⓑ | ⓒ | ⓓ |
| 14. | ⓐ | ⓑ | ⓒ | ⓓ |
| 15. | ⓐ | ⓑ | ⓒ | ⓓ |
| 16. | ⓐ | ⓑ | ⓒ | ⓓ |
| 17. | ⓐ | ⓑ | ⓒ | ⓓ |
| 18. | ⓐ | ⓑ | ⓒ | ⓓ |
| 19. | ⓐ | ⓑ | ⓒ | ⓓ |
| 20. | ⓐ | ⓑ | ⓒ | ⓓ |
| 21. | ⓐ | ⓑ | ⓒ | ⓓ |
| 22. | ⓐ | ⓑ | ⓒ | ⓓ |
| 23. | ⓐ | ⓑ | ⓒ | ⓓ |
| 24. | ⓐ | ⓑ | ⓒ | ⓓ |
| 25. | ⓐ | ⓑ | ⓒ | ⓓ |
| 26. | ⓐ | ⓑ | ⓒ | ⓓ |
| 27. | ⓐ | ⓑ | ⓒ | ⓓ |
| 28. | ⓐ | ⓑ | ⓒ | ⓓ |
| 29. | ⓐ | ⓑ | ⓒ | ⓓ |
| 30. | ⓐ | ⓑ | ⓒ | ⓓ |
| 31. | ⓐ | ⓑ | ⓒ | ⓓ |
| 32. | ⓐ | ⓑ | ⓒ | ⓓ |
| 33. | ⓐ | ⓑ | ⓒ | ⓓ |
| 34. | ⓐ | ⓑ | ⓒ | ⓓ |
| 35. | ⓐ | ⓑ | ⓒ | ⓓ |
| 36. | ⓐ | ⓑ | ⓒ | ⓓ |
| 37. | ⓐ | ⓑ | ⓒ | ⓓ |
| 38. | ⓐ | ⓑ | ⓒ | ⓓ |
| 39. | ⓐ | ⓑ | ⓒ | ⓓ |
| 40. | ⓐ | ⓑ | ⓒ | ⓓ |
| 41. | ⓐ | ⓑ | ⓒ | ⓓ |
| 42. | ⓐ | ⓑ | ⓒ | ⓓ |
| 43. | ⓐ | ⓑ | ⓒ | ⓓ |
| 44. | ⓐ | ⓑ | ⓒ | ⓓ |
| 45. | ⓐ | ⓑ | ⓒ | ⓓ |
| 46. | ⓐ | ⓑ | ⓒ | ⓓ |
| 47. | ⓐ | ⓑ | ⓒ | ⓓ |
| 48. | ⓐ | ⓑ | ⓒ | ⓓ |
| 49. | ⓐ | ⓑ | ⓒ | ⓓ |
| 50. | ⓐ | ⓑ | ⓒ | ⓓ |

*(handwritten top margin)*
$16$
$\times 3$
$48$

**1.** What is the value of $f(4)$ when
$f(x) = 3x^2 - \sqrt{x}$?

*(handwritten)* $= 3(4)^2 - \sqrt{4}$
$= 3 \cdot 16 - 2$
$= 48 - 2$
$= 46$

  a.  44
  b.  46
  c.  140
  d.  142

**2.** Simplify $g(x + 3)$ when $g(x) = x^2 - 2x + 1$.

  a.  $x^2 + 4x + 4$
  b.  $x^2 - 2x + 4$
  c.  $x^2 - 2x + 13$
  d.  $x^2 + 4x + 10$

*(handwritten)* $(x+3)^2 - 2(x+3)+1$
$x^2 + 6x + 9 - 2x - 6 + 1$
$x^2 + 4x + 10$

**3.** What is $f \circ g(x)$ when $f(x) = x - \dfrac{2}{x}$ and
$g(x) = x + 3$?

  a.  $x - \dfrac{2}{x} + 3$
  b.  $2x - \dfrac{2}{x} + 3$
  c.  $x^2 - 2 + 3x - \dfrac{6}{x}$
  d.  $x + 3 - \dfrac{2}{x + 3}$

*(handwritten)* $\left(x - \dfrac{2}{x}\right) \cdot (x+3)$
$x^2 + 3x - 2 - \dfrac{6}{x}$
$x^2 + 3x - 2 - \dfrac{6}{x}$

**4.** What is the domain of $h(x) = \dfrac{x}{x^2 - 1}$?

  a.  $x \neq 1$
  b.  $x \neq 0$
  c.  $x \neq -1$ and $x \neq 1$
  d.  $x \neq -1$, $x \neq 0$, and $x \neq 1$

Use the following figure for questions 5 and 6.

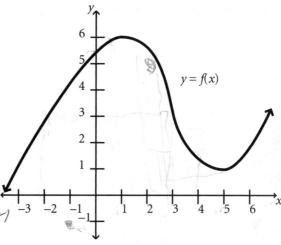

$y = f(x)$

**5.** Where is $f(x)$ increasing?

  a.  $(-\infty, 1)$ and $(5, \infty)$
  b.  $(1, 5)$
  c.  $(1, 6)$
  d.  $(5, \infty)$

**6.** Where does $f(x)$ have a point of inflection?

  a.  $(0, 5.5)$
  b.  $(1, 6)$
  c.  $(3, 3)$
  d.  $(5, 1)$

**7.** What is the equation of the straight line through
$(2, 5)$ and $(-1, -1)$?

  a.  $y = 2x + 5$
  b.  $y = 2x + 1$
  c.  $y = -2x + 9$
  d.  $y = -2x - 3$

*(handwritten)*
$y = mx + b$
$m = \dfrac{y_2 - y_1}{x_2 - x_1}$
$= \dfrac{5 + 1}{2 + 1} = \dfrac{6}{3}$
$= 2$

$1 = 2(-1) + b$
$-1 = -2 + b$
$-1 + 2 = b$
$1 = b$

**8.** Simplify $64^{\frac{1}{2}}$.

  **a.** 4
  **b.** 8
  **c.** 32
  **d.** 4,096

**9.** Simplify $2^{-3}$.

  **a.** $\frac{1}{8}$
  **b.** 8
  **c.** $-8$
  **d.** $-6$

**10.** Solve for $x$ when $3^x = 15$.

  **a.** 5
  **b.** $\ln(5)$
  **c.** $\frac{\ln(15)}{\ln(3)}$
  **d.** $\ln(12)$

**11.** Evaluate $\sin\left(\frac{\pi}{3}\right)$.

  **a.** $-\frac{1}{2}$
  **b.** $\frac{1}{2}$
  **c.** $\frac{\sqrt{2}}{2}$
  **d.** $\frac{\sqrt{3}}{2}$

**12.** Evaluate $\tan\left(\frac{3\pi}{4}\right)$.

  **a.** $-1$
  **b.** 1
  **c.** $\frac{\sqrt{2}}{2}$
  **d.** $\sqrt{2}$

**13.** Simplify $\lim\limits_{x\to 4}\frac{x^2-1}{x^2+1}$.

  **a.** $-1$
  **b.** $\frac{3}{5}$
  **c.** $\frac{15}{17}$
  **d.** $\frac{7}{9}$

**14.** Simplify $\lim\limits_{x\to 1}\frac{x-1}{x^2-1}$.

  **a.** $-1$
  **b.** 1
  **c.** $\frac{1}{2}$
  **d.** undefined

**15.** Evaluate $\lim\limits_{x\to 2^-}\frac{x+3}{x-2}$.

  **a.** $\infty$
  **b.** $-\infty$
  **c.** $-\frac{1}{4}$
  **d.** undefined

**16.** What is the slope of $f(x) = 3x + 2$ at $x = 5$?

   **a.** 2

   **b.** 17

   **c.** $3x$

   **d.** 3

**17.** What is the slope of $g(x) = x^2 + 2x - 1$ at $x = 3$?

   **a.** 2

   **b.** 8

   **c.** 14

   **d.** $2x + 2$

**18.** Differentiate $h(x) = 4x^3 - 5x + 1$.

   **a.** $12x^2$

   **b.** $12x^2 - 5$

   **c.** $12x^2 - 5x$

   **d.** $12x^2 - 5x + \dfrac{1}{x}$

**19.** The height of a certain plant is $41 - \dfrac{40}{t}$ inches after $t \geq 1$ weeks. How fast is it growing after two weeks?

   **a.** 5 inches per week

   **b.** 10 inches per week

   **c.** 21 inches per week

   **d.** 31 inches per week

**20.** What is the derivative of $y = x^2 - 3\cos(x)$?

   **a.** $\dfrac{dy}{dx} = 2x + 3\sin(x)$

   **b.** $\dfrac{dy}{dx} = 2x - 3\sin(x)$

   **c.** $\dfrac{dy}{dx} = 2x - 3\cos(1)$

   **d.** $\dfrac{dy}{dx} = 2x - 3\tan(x)$

**21.** Differentiate $f(x) = \ln(x) - e^x + 2$.

   **a.** $f'(x) = \ln(x) + e^x$

   **b.** $f'(x) = \ln(x) - e^x$

   **c.** $f'(x) = \dfrac{1}{x} + e^x$

   **d.** $f'(x) = \dfrac{1}{x} - e^x$

**22.** Differentiate $g(x) = x^2\sin(x)$.

   **a.** $2x\cos(x)$

   **b.** $2x + \cos(x)$

   **c.** $2x\sin(x) + x^2\cos(x)$

   **d.** $2x\sin(x)\cos(x)$

**23.** Differentiate $\dfrac{\ln(x)}{x}$.

  **a.** 0

  **b.** $\dfrac{1}{x}$

  **c.** $\dfrac{1 - \ln(x)}{x^2}$

  **d.** $\dfrac{\ln(x) - 1}{x^2}$

**24.** Differentiate $y = \tan(x)$.

  **a.** $\sec^2(x)$

  **b.** $\csc(x)$

  **c.** $\dfrac{\cos^2(x) - \sin^2(x)}{\cos^2(x)}$

  **d.** $\sin(x)\cos(x)$

**25.** Differentiate $f(x) = e^{4x^2 + 7}$.

  **a.** $e^{8x}$

  **b.** $e^{4x^2 + 7}$

  **c.** $8xe^{4x^2 + 7}$

  **d.** $(4x^2 + 7)e^{4x^2 - 8}$

**26.** Differentiate $(x^2 - 1)^5$.

  **a.** $10x$

  **b.** $(2x)^5$

  **c.** $5(x^2 - 1)^4$

  **d.** $10x(x^2 - 1)^4$

**27.** Find $\dfrac{dy}{dx}$ if $y^2 + xy = x^3 + 5$.

  **a.** $\dfrac{dy}{dx} = x^2$

  **b.** $\dfrac{dy}{dx} = \dfrac{3x^2 - y}{2y + x}$

  **c.** $\dfrac{dy}{dx} = \dfrac{3x^2}{1 + 2y}$

  **d.** $\dfrac{dy}{dx} = \dfrac{3x^2 - 3y}{x}$

**28.** Find $\dfrac{dy}{dx}$ if $\sin(y) = 4x^2$.

  **a.** $\dfrac{dy}{dx} = 8x - \cos(y)$

  **b.** $\dfrac{dy}{dx} = 8x\cos(y)$

  **c.** $\dfrac{dy}{dx} = \cos(y) - 8x$

  **d.** $\dfrac{dy}{dx} = 8x\sec(y)$

**29.** What is the slope of $x^2 + y^2 = 1$ at $\left(\dfrac{1}{2}, \dfrac{\sqrt{3}}{2}\right)$?

a. $-1$

b. $1$

c. $-\dfrac{\sqrt{3}}{3}$

d. $\dfrac{\sqrt{3}}{3}$

**30.** If the radius of a circle is growing at 4 feet per second, how fast is the area growing when the radius is 10 feet?

a. $20\pi$ square feet per second

b. $80\pi$ square feet per second

c. $100\pi$ square feet per second

d. $400\pi$ square feet per second

**31.** The height of a triangle increases by 3 inches every minute while its base decreases by 1 inch every minute. How fast is the area changing when the triangle has a height of 10 inches and a base of 100 inches?

a. It is increasing at 145 square inches per minute.

b. It is increasing at 500 square inches per minute.

c. It is decreasing at 1,500 square inches per minute.

d. It is decreasing at 3,000 square inches per minute.

**32.** Evaluate $\displaystyle\lim_{x \to \infty} \dfrac{4x^2 - 5x + 2}{1 - x^2}$.

a. $4$

b. $-4$

c. $2$

d. undefined

**33.** Evaluate $\displaystyle\lim_{x \to -\infty} \dfrac{4x^5 + 6x + 4}{x^3 + 10x - 1}$.

a. $-\infty$

b. $\infty$

c. $-4$

d. $4$

**34.** Evaluate $\displaystyle\lim_{x \to \infty} \dfrac{\ln(x)}{3x + 2}$.

a. $\dfrac{1}{3}$

b. $2$

c. $3$

d. $0$

**35.** Which of the following is the graph of
$y = \dfrac{1}{x - 2}$?

**a.**

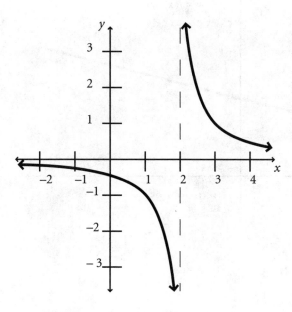

**b.**

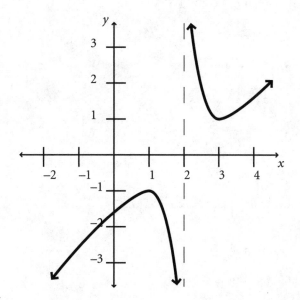

**c.**

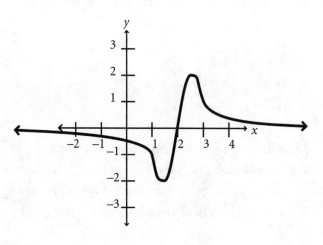

**d.**

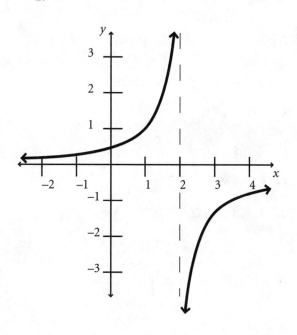

**36.** Where is $g(x) = x^4 - 6x^2 + 5$ concave down?

   **a.** $(1,12)$

   **b.** $(-6,5)$

   **c.** $(-\sqrt{3}, \sqrt{3})$

   **d.** $(-1,1)$

**37.** A 5,000-pound block of ice melts at a rate of 200 pounds each day. If the price of ice is 5¢ a pound and increases by 1¢ each day, in how many days will the block have a maximal value?

   **a.** 5 days

   **b.** 10 days

   **c.** 15 days

   **d.** 20 days

**38.** A box with a square bottom and no top must contain 108 cubic inches. What dimensions will minimize the surface area of the box?

   **a.** $2 \times 2 \times 27$

   **b.** $8 \times 8 \times 3$

   **c.** $6 \times 6 \times 3$

   **d.** $4 \times 4 \times 6.75$

**39.** If $\int_3^8 g(x)\,dx = 5$ and $\int_3^5 g(x)\,dx = -4$, then what is $\int_5^8 g(x)\,dx$?

   **a.** $-20$

   **b.** $1$

   **c.** $3$

   **d.** $9$

**40.** What is $\int_0^4 f(x)\,dx$?

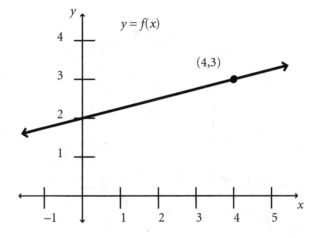

   **a.** 2

   **b.** 3

   **c.** 10

   **d.** 12

**41.** If $g(x)$ is the area under the curve $y = t^3 + 4t$ between $t = 0$ and $t = x$, what is $g'(x)$?

   **a.** $x^3 + 4x$

   **b.** $3x^2 + 4$

   **c.** $\frac{1}{4}x^4 + 2x$

   **d.** $0$

**42.** Evaluate $\int (3x^2 + 8x + 5)\,dx$.

   **a.** $6x + 8$

   **b.** $6x + 8 + c$

   **c.** $x^3 + 4x^2 + 5x$

   **d.** $x^3 + 4x^2 + 5x + c$

**43.** Evaluate $\int_0^9 \sqrt{x}\,dx$.

    **a.** $\dfrac{1}{6}$

    **b.** 3

    **c.** 12

    **d.** 18

**44.** Evaluate $\int \sin(x)\,dx$.

    **a.** $\cos(x) + c$

    **b.** $-\cos(x) + c$

    **c.** $\sin(x) + c$

    **d.** $-\sin(x) + c$

**45.** Evaluate $\int \dfrac{x}{x^2 - 1}\,dx$.

    **a.** $\dfrac{\frac{1}{2}x^2}{\frac{1}{3}x^3 - x} + c$

    **b.** $\ln(x - 1) + c$

    **c.** $\ln(x^2 - 1) + c$

    **d.** $\dfrac{1}{2}\ln(x^2 - 1) + c$

**46.** Integrate $\int e^{5x}\,dx$.

    **a.** $\dfrac{1}{5}e^{5x} + c$

    **b.** $e^{5x} + c$

    **c.** $e^5 + c$

    **d.** $\dfrac{1}{5}e^5 + c$

**47.** Evaluate $\int 4x^2\cos(x^3)\,dx$.

    **a.** $4\sin(x^3) + c$

    **b.** $\dfrac{4}{3}\sin(x^3) + c$

    **c.** $\dfrac{4}{3}x^3\sin(x^3) + c$

    **d.** $\dfrac{4}{3}x^2\sin(x^3) + c$

**48.** Evaluate $\int_0^1 x(x^2 + 2)^5\,dx$.

    **a.** 73

    **b.** 81

    **c.** $\dfrac{665}{12}$

    **d.** $\dfrac{81}{8}$

**49.** Integrate $\int x\ln(x)\,dx$.

   **a.** $\frac{1}{2}x^2\ln(x) + c$

   **b.** $x\ln(x) - \ln(x) + c$

   **c.** $x^2\ln(x) + \frac{1}{4}x^2 + c$

   **d.** $\frac{1}{2}x^2\ln(x) - \frac{1}{4}x^2 + c$

**50.** Evaluate $\int x\sin(x)\,dx$.

   **a.** $-x\cos(x) + \sin(x) + c$

   **b.** $\frac{1}{2}x^2\cos(x) + c$

   **c.** $-\frac{1}{2}x^2\cos(x) + c$

   **d.** $-x\cos(x) + \cos(x) + c$

## ▶ Answers

1. b. Lesson 1
2. a. Lesson 1
3. d. Lesson 1
4. c. Lesson 1
5. a. Lesson 2
6. c. Lesson 2
7. b. Lesson 2
8. b. Lesson 3
9. a. Lesson 3
10. c. Lesson 3
11. d. Lesson 4
12. a. Lesson 4
13. c. Lesson 5
14. c. Lesson 5
15. b. Lesson 5
16. d. Lessons 6, 7
17. b. Lessons 6, 7
18. b. Lesson 7
19. b. Lesson 8
20. a. Lesson 8
21. d. Lesson 8
22. c. Lesson 9
23. c. Lessons 8, 9
24. a. Lesson 9
25. c. Lesson 10
26. d. Lesson 10
27. b. Lesson 11
28. d. Lessons 4, 11
29. c. Lesson 11
30. b. Lesson 12
31. a. Lesson 12
32. b. Lesson 13
33. b. Lesson 13
34. d. Lesson 13
35. a. Lesson 14
36. d. Lesson 14
37. b. Lesson 15
38. c. Lesson 16
39. d. Lesson 16
40. c. Lesson 16
41. a. Lesson 17
42. d. Lesson 18
43. d. Lesson 18
44. b. Lesson 18
45. d. Lesson 19
46. a. Lesson 19
47. b. Lesson 19
48. c. Lesson 19
49. d. Lesson 20
50. a. Lesson 20

# 1 ▶ Functions

Calculus is the study of change. It is often important to know when something is increasing, when it is decreasing, and when it hits a high or low point. Much of the business of finance depends on predicting the high and low points for prices. In science and engineering, it is often essential to know precisely how fast quantities such as temperature, size, and speed are changing. Calculus is the primary tool for calculating such changes.

Numbers, which are the focus of arithmetic, are no longer the objects of our study. This is because they do not change. The number 5 will always be 5. It never goes up or down. Thus, we need to introduce a new sort of mathematical object, something that *can* change. These objects, the centerpiece of calculus, are functions.

## ▶ Functions

A *function* is a way of matching up one set of numbers with another. The first set of numbers is called the *domain*. For each of these numbers in a set, the function assigns exactly one number from the other set, the *range.*

It is true that in algebra, everyone is taught "parentheses mean multiplication." This means that $5(2 + 7) = 5(9) = 45$. If $x$ is a variable, then $x(2 + 7) = x(9) = 9x$. However, if $f$ is the name of a function, then $f(2 + 7) = f(9) =$ the number to which $f$ takes 9. The expression $f(x)$ is pronounced "$f$ of $x$" and not "$f$ times $x$." This can be confusing, so an apology is necessary. Mathematicians use parentheses to mean several different things and expect everyone to know the difference. Sorry!

For example, the domain of the function could be the numbers 1, 4, 9, 25, and 100; and the range could be 1, 2, 3, 5, and 10. Suppose the function takes 1 to 1, 4 to 2, 9 to 3, 25 to 5, and 100 to 10. This could be illustrated by the following:

$$1 \to 1$$
$$4 \to 2$$
$$9 \to 3$$
$$25 \to 5$$
$$100 \to 10$$

Because we sometimes use several functions at the same time, we give them names. Let us call the function we just mentioned by the name *Eugene*. Thus, we can ask, "Hey, what does Eugene do with the number 4?" The answer is "Eugene takes 4 to the number 2."

Mathematicians are notoriously lazy, so we try to do as little writing as possible. Thus, instead of writing "Eugene takes 4 to the number 2," we often write "Eugene(4) = 2" to mean the same thing. Similarly, we like to use names that are as short as possible, such as $f$ (for function), $g$ (for function when $f$ is already being used), $h$, and so on. The trigonometric functions in Lesson 4 all have three-letter names like sin and cos, but even these are abbreviations. So let us save space and use $f$ instead of Eugene.

Because the domain is small, it is easy to write out everything:

$$f(1) = 1$$
$$f(4) = 2$$
$$f(9) = 3$$
$$f(25) = 5$$
$$f(100) = 10$$

However, if the domain were large, this would get very tedious. It is much easier to find a pattern and use that pattern to describe the function. Our function $f$ just happens to take each number of its domain to the square root of that number. Therefore, we can describe $f$ by saying:

$f$(a number) = the square root of that number

Of course, anyone with experience in algebra knows that writing "a number" over and over is a waste of time. Why not just pick a *variable* to represent the number? Just as $f$ is our favorite name for functions, little $x$ is the most beloved of all variable names. Here is the way to represent our function $f$ with the absolute least amount of writing necessary:

$$f(x) = \sqrt{x}$$

This tells us that putting a number into the function $f$ is the same as putting it into $\sqrt{\ }$. Thus,

$$f(25) = \sqrt{25} = 5 \text{ and } f(4) = \sqrt{4} = 2.$$

### Example
Find the value of $g(3)$ if $g(x) = x^2 + 2$.

### Solution

Replace each occurrence of $x$ with 3.

$g(3) = 3^2 + 2$

Simplify.

$g(3) = 9 + 2 = 11$

### Example

Find the value of $h(-2)$ if $h(t) = t^3 - 2t^2 + 5$.

### Solution

Replace each occurrence of $t$ with $-2$.

$h(-2) = (-2)^3 - 2(-2)^2 + 5$

Simplify.

$h(-2) = -8 - 2(4) + 5 = -8 - 8 + 5 = -11$

### ▶ Practice

**1.** Find the value of $f(5)$ when $f(x) = 2x - 1$.

**2.** Find the value of $g(-3)$ when
$g(x) = x^3 + x^2 + x + 1$.

**3.** Find the value of $h\left(\dfrac{1}{2}\right)$ when $h(t) = t^2 + \dfrac{3}{4}$.

**4.** Find the value of $f(7)$ when $f(x) = 2$.

**5.** Find the value of $k(4)$ when
$k(u) = u^2 + 2u - \dfrac{12}{u}$.

**6.** Find the value of $h(64)$ when
$h(x) = \sqrt{x} - \sqrt[3]{x}$.

**7.** Suppose that after $t$ seconds, a rock thrown off a bridge has height $s(t) = -16t^2 + 20t + 100$ feet off the ground. How high is it after 3 seconds?

**8.** Suppose that the profit on making and selling $x$ cookies is $P(x) = \dfrac{x}{2} - \dfrac{x^2}{10{,}000} - \$10$. How much profit is made on 100 cookies?

### ▶ Plugging Variables into Functions

Variables can be plugged into functions just as easily as numbers can. Often, though, they can't be simplified as much.

### Example

Simplify $f(w)$ if $f(x) = \sqrt{x} + 2x^2 + 2$.

### Solution

Replace each occurrence of $x$ with $w$.

$f(w) = \sqrt{w} + 2w^2 + 2$

That is all we can say without knowing more about $w$.

### Example

Simplify $g(a + 5)$ if $g(t) = t^2 - 3t + 1$.

### Solution

Replace each occurrence of $t$ with $(a + 5)$.

$g(a + 5) = (a + 5)^2 - 3(a + 5) + 1$

Multiply out $(a + 5)^2$ and $-3(a + 5)$.

$g(a + 5) = a^2 + 10a + 25 - 3a - 15 + 1$

Simplify.

$g(a + 5) = a^2 + 7a + 11$

## Example

Simplify $\dfrac{f(x + a) - f(x)}{a}$ if $f(x) = x^2$.

## Solution

Start with what needs to be simplified.

$$\frac{f(x + a) - f(x)}{a}$$

Use $f(x) = x^2$ to evaluate $f(x + a)$ and $f(x)$.

$$\frac{(x + a)^2 - x^2}{a}$$

Multiply out $(x + a)^2$.

$$\frac{x^2 + 2xa + a^2 - x^2}{a}$$

Cancel out the $x^2$ and the $-x^2$.

$$\frac{2xa + a^2}{a}$$

Factor out an $a$.

$$\frac{(2x + a)a}{a}$$

Cancel an $a$ from the top and bottom.

$$2x + a$$

## ▶ Practice

Simplify the following.

**9.** $f(y)$ when $f(x) = x^2 + 3x - 1$

**10.** $f(y + 1)$ when $f(x) = x^2 + 3x - 1$

**11.** $f(x + a)$ when $f(x) = x^2 + 3x - 1$

**12.** $g(x^2 + \sqrt{x})$ when $g(t) = \dfrac{8}{t} - 6t$

**13.** $g(2x) - g(x)$ when $g(t) = \dfrac{8}{t} - 6t$

**14.** $f(x + a) - f(x)$ when $f(x) = x^2 + 4x - 5$

**15.** $\dfrac{h(x + a) - h(x)}{a}$ when $h(x) = 3x + 2$

**16.** $\dfrac{g(x + a) - g(x)}{a}$ when $g(x) = x^2 - 2x + 1$

## ▶ Composition

Now that we can plug anything into functions, we can plug one function into another. This is called *composition*. The composition of function $f$ with function $g$ is written $f \circ g$. This means to plug $g$ into $f$ like this:

$$f \circ g(x) = f(g(x))$$

It may seem that $f$ comes first in $f \circ g(x)$, reading from left to right, but actually, the $g$ is closer to the $x$. This means that the function $g$ acts on the $x$ first.

## Example

If $f(x) = \sqrt{x} + 2x$ and $g(x) = 4x + 7$, then what is the composition $f \circ g(x)$?

## Solution

Start with the definition of composition.

$$f \circ g(x) = f(g(x))$$

Use $g(x) = 4x + 7$.

$$f \circ g(x) = f(4x + 7)$$

Replace each occurrence of $x$ in $f$ with $4x + 7$.

$$f \circ g(x) = \sqrt{4x + 7} + 2(4x + 7)$$

Simplify.

$$f \circ g(x) = \sqrt{4x + 7} + 8x + 14$$

Conversely, to evaluate $g \circ f(x)$, we compute:

$$g \circ f(x) = g(f(x))$$

Use $f(x) = \sqrt{x} + 2x$.

$$g \circ f(x) = g(\sqrt{x} + 2x)$$

Replace each occurrence of $x$ in $g$ with $\sqrt{x} + 2x$.

$$g \circ f(x) = 4(\sqrt{x} + 2x) + 7$$

Simplify.

$$g \circ f(x) = 4\sqrt{x} + 8x + 7$$

Notice that $f \circ g(x)$ and $g \circ f(x)$ are different. This is usually the case.

### Example

If $f(x) = x^2 + 2x + 1$ and $g(x) = 5x + 1$, then what is $f \circ g(x)$?

### Solution

Start with the definition of composition.

$$f \circ g(x) = f(g(x))$$

Use $g(x) = 5x + 1$.

$$f \circ g(x) = f(5x + 1)$$

Replace each occurrence of $x$ in $f$ with $5x + 1$.

$$f \circ g(x) = (5x + 1)^2 + 2(5x + 1) + 1$$

Simplify.

$$f \circ g(x) = 25x^2 + 20x + 4$$

## ▶ Practice

Using $f(x) = \dfrac{1}{x}$, $g(x) = x^3 - 2x^2 + 1$, and $h(x) = x - \sqrt{x}$, simplify the following compositions.

**17.** $f \circ g(x)$

**18.** $g \circ f(x)$

**19.** $f \circ h(t)$

**20.** $f \circ f(x)$

**21.** $h \circ h(x)$

**22.** $g \circ h(9)$

**23.** $h \circ f \circ g(x)$

**24.** $f \circ h \circ f(2x)$

## ▶ Domains

In the beginning of the lesson, we defined the function Eugene as:

$$f(x) = \sqrt{x}$$

However, we left out a crucial piece of information: the domain. The domain of this function consisted of only the numbers 1, 4, 9, 25, and 100. Thus, we should have written

$$f(x) = \sqrt{x} \text{ if } x = 1, 4, 9, 25, \text{ or } 100$$

Usually, the domain of a function is not given explicitly like this. In such situations, it is assumed that the domain is as large as it possibly can be. The domain consists of all numbers that don't violate one of the following two fundamental prohibitions:

- Never divide by zero.
- Never take an even root of a negative number.

If you divide by zero, the entire numerical universe will collapse down to a single point. If dividing by zero were allowed, then all numbers would be equal. Four would equal five. Negative and positive would be equivalent. "It's all the same to me" would be the correct answer to every math question. While this might be appealing to some people, it would make calculus, the study of change, impossible. If only one number existed, there could be no change. Thus, we automatically rule out any situation where division by zero might occur.

### Example

What is the domain of $f(x) = \dfrac{3}{x - 2}$?

### Solution

We must never let the denominator $x - 2$ be zero, so $x$ cannot be 2. Therefore, the domain of this function consists of all real numbers except 2.

The prohibition against even roots (like square roots) of negative numbers is less severe. An even root of a negative number is an imaginary number. Useful mathematics can be done with imaginary numbers. However, for the sake of simplicity, we will avoid them in this book.

### Example

What is the domain of $g(x) = \sqrt{3x + 2}$?

### Solution

The numbers in the square root must not be negative, so $3x + 2 \geq 0$, thus $x \geq -\dfrac{2}{3}$. The domain consists of all numbers greater than or equal to $-\dfrac{2}{3}$.

Do note that it is perfectly okay to take the square root of zero, since $\sqrt{0} = 0$. It is only when numbers are less than zero that even roots become imaginary.

### Example

Find the domain of $k(x) = \dfrac{\sqrt{4 - x}}{x^2 + 5x + 6}$.

### Solution

To avoid dividing by zero, we need $x^2 + 5x + 6 \neq 0$, so $(x + 3)(x + 2) \neq 0$, thus $x \neq -3$ and $x \neq -2$. To avoid an even root of a negative number, $4 - x \geq 0$, so $x \leq 4$. Thus, the domain of $k$ is $x \leq 4$, $x \neq -3$, $x \neq -2$.

## ▶ Practice

Find the domain of each of the following functions.

**25.** $f(x) = \dfrac{1}{(x + 3)(x - 5)}$

**26.** $h(x) = \sqrt{x + 1}$

**27.** $k(t) = \dfrac{1}{\sqrt{t + 5}}$

**28.** $g(x) = x^2 + 5x - 6$

**29.** $f(a) = \dfrac{3x + 7}{a}$

**30.** $h(x) = \sqrt[3]{x}$

**31.** $k(x) = \dfrac{\sqrt[4]{2 - x}}{x + 8}$

**32.** $g(u) = \dfrac{8u}{\sqrt{4 + 3u}(u + 3)}$

# LESSON

# 2 ▶ Graphs

A function can be fully described by showing what happens at each number in its domain (for example, $4 \rightarrow 2$) or by giving its formula (for example, $f(x) = \sqrt{x}$). However, neither of these provides a clear overall picture of the function.

Luckily for us, René Descartes came up with the idea of a *graph*, a visual picture of a function. Rather than say $4 \rightarrow 2$ or $f(4) = 2$, we plot $(4,2)$ on the Cartesian plane, which would look like Figure 2.1.

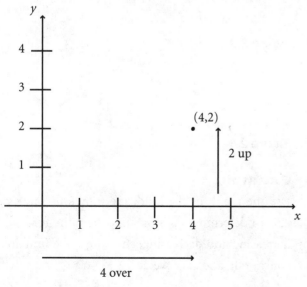

**Figure 2.1**

We put the $y$ into the formula $y = f(x) = \sqrt{x}$ to imply that the $y$-coordinates of our points are the numbers we get by plugging the $x$-coordinates into the function $f$.

## ▶ Practice

Plot the following points on a Cartesian plane.

**1.** $(3,5)$

**2.** $(-3,4)$

**3.** $(2,-6)$

**4.** $(-1,-5)$

**5.** $(0,3)$

**6.** $(-5,0)$

**7.** $(0,0)$

**8.** $\left(4\frac{1}{2}, \frac{1}{4}\right)$

For the function $f(x) = x^2 - 2x + 5$, plot the point at the following positions.

**9.** $x = 3$

**10.** $x = 1$

**11.** $x = 0$

**12.** $x = -2$

If we plotted *all* the points in the domain of $f(x) = \sqrt{x}$ (not just the whole numbers, but all the fractions and decimals, too), then the points would be so close together that they would form a continuous curve as in Figure 2.2.

The graph shows us several interesting characteristics of the function $f(x) = \sqrt{x}$. Because the graph starts at $x = 0$ and runs to the right, this means that the domain is $x \geq 0$.

We can see that the function $f(x) = \sqrt{x}$ is *increasing* (going up from left to right) and not *decreasing* (going down from left to right).

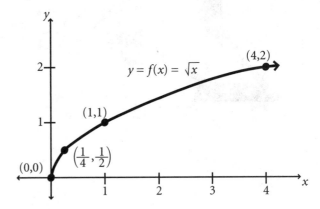

**Figure 2.2**

The function $f(x) = \sqrt{x}$ is *concave down* because it curves downward (see Figure 2.3) like a frown and not *concave up* like a smile (see Figure 2.4).

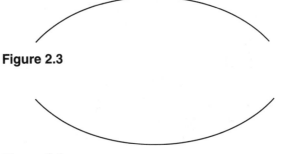

**Figure 2.3**

**Figure 2.4**

## Example

Use the graph of the following function (see Figure 2.5) to determine the domain, where the function is increasing and decreasing, and where the function is concave up and concave down.

An apology must be made for mathematical notation here. An expression like (2,8) is ambiguous. Is this a single point with coordinates $x = 2$ and $y = 8$? Is this an interval consisting of all the points between 2 and 8? Only the context can make clear which is meant. If we read "at (2,8)," then this is a single point. If we read "on (2,8)," then it refers to an interval.

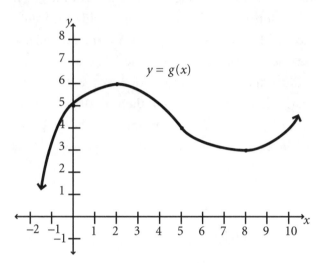

**Figure 2.5**

## Solution

The domain of $g$ consists of all real numbers because there is a point above or below every number on the $x$-axis.

The function $g$ is increasing up to the point at $x = 2$, where it then decreases down to $x = 8$, and then increases ever afterward. To save space, we say that $g$ increases on $(-\infty, 2)$ and on $(8, \infty)$, and that $g$ decreases on $(2,8)$.

The point at $(2,6)$ where $g$ stops increasing and begins to decrease is the highest point in its immediate area and is called a *local maximum*. The point at $(8,3)$ is similarly a *local minimum*, the lowest point in its neighborhood. These points tend to be the most interesting points on a graph.

The concavity of $g$ is trickier to estimate. Clearly $g$ is concave down in the vicinity of $x = 2$ and concave

up around $x = 7$ and $x = 8$. The exact point where the concavity changes is called a *point of inflection*. On this graph, it seems to be at the point $(5,4)$, though some people might imagine it a bit earlier or later. Thus, we say that $g$ is concave down on $(-\infty, 5)$ and concave up on $(5, \infty)$.

To be completely honest, any information obtained by looking at a graph is going to be a rough estimate. Is the local maximum at $(2,6)$, or is it at $(2.0003, 5.9998)$? There is no way to tell the difference. Graphs made up by people, like the ones in this lesson, tend to have everything interesting happen at whole numbers. Graphs formed using real-world data tend to be much less kind.

## Example

Use the graph in Figure 2.6 to identify the domain of $h$, where it is increasing and decreasing, where it has local maxima and minima, where it is concave up and down, and where it has points of inflection.

## Solution

The first thing to notice is that $h$ has three breaks, or *discontinuities*. If we wanted to trace the graph of $h$ with a continuous motion of a pencil, then we would have to lift up the pencil at $x = -2$, $x = 2$, and at $x = 5$. The little circle at $(5,3)$ indicates a hole in the graph where a single point has been taken out. This means that $x = 5$ is not in the domain, just as $x = -1$ has no point above or below it. The situation at $x = 2$ is more interesting because $x = 2$ *is* in the domain,

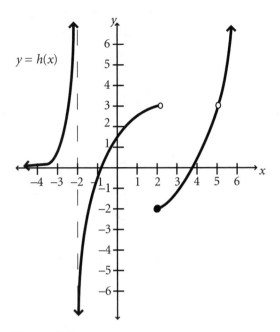

**Figure 2.6**

because the $y$-values get really close to $y = 3$; there is no highest point in the range.

Similarly, a point of inflection can be seen at $x = 2$ but not at $x = -2$ because there can't be a point of inflection where there is no point!

The situation at $x = -2$ is called an *asymptote* because the graph begins to flatten out like a straight line. The more we would continue to draw the graph off the top and bottom, the straighter this line would become. In this case, $x = -2$ is a *vertical asymptote* because it approximates a vertical line at $x = -2$. Because the graph appears to flatten out like the straight horizontal line $y = 0$ (the $x$-axis) as the graph goes off to the left, this means that the graph of $y = h(x)$ appears to have a *horizontal asymptote* at $y = 0$.

with the point (the shaded-in circle) at $(2,-2)$ representing $h(2) = -2$. All of the points immediately before $x = 2$ have $y$-values close to $y = 3$, but then there is an abrupt jump down to $x = 2$. Such jumps look awkward on a graph, but occur often in real life, like the way the cost of postage leaps up as soon as a letter weighs more than one ounce.

Because of the discontinuities, we have to name each interval separately, as in: $h$ increases on $(-\infty,-2)$, $(-2,2)$, $(2,5)$, and on $(5,\infty)$. As well, $h$ is concave up on $(-\infty,-2)$, $(2,5)$, and on $(5,\infty)$, and concave down on $(-2,2)$.

There is a local minimum at $(2,-2)$, because the point there is the lowest in its immediate vicinity, $1 < x < 3$. There is no local maximum in that range

## ▶ Practice

Use the graph of each function to determine the domain, the discontinuities, where the function is increasing and decreasing, the local maximum and minimum points, where the function is concave up and down, the points of inflection, and the asymptotes.

**13.**

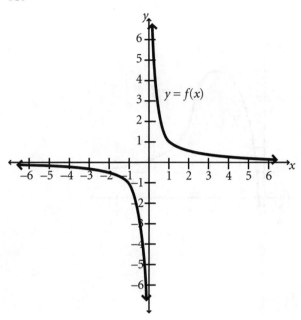

**15.**

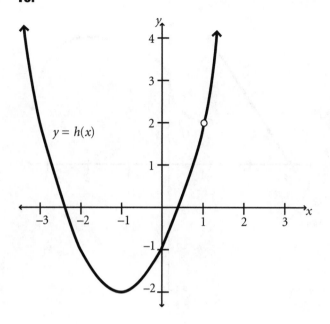

**14.**

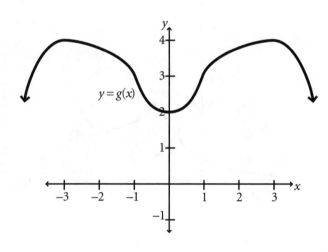

**16.**

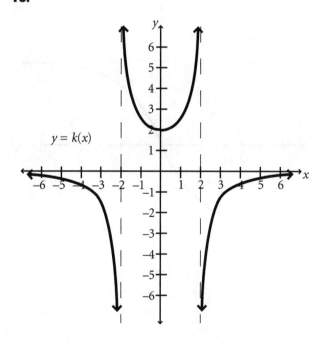

**17.**

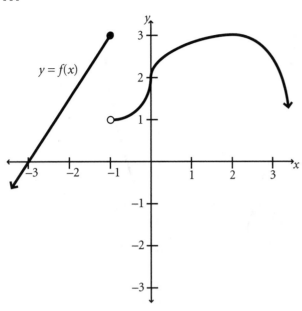

**19.**

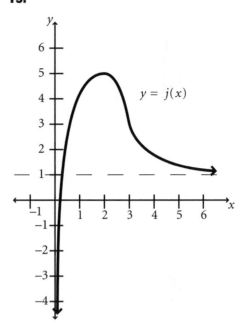

**18.**

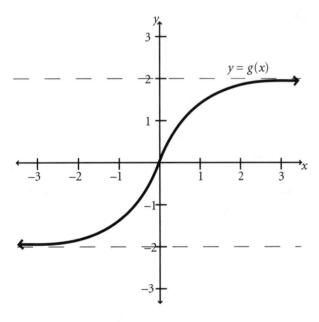

**20.**

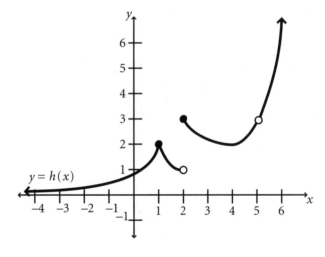

### ▶ Note

We can obtain all sorts of useful information from a graph, such as its maximal points, where it is increasing and decreasing, and so on. Calculus will enable us to get this information directly from the function. We will then be able to draw graphs intelligently, without having to calculate and plot thousands of points (the method graphing calculators use).

## ▶ Straight Lines

The easiest and most beloved of all graphs are straight lines. Human beings tend to build, move, and even think in straight lines. There is something calming and reassuring about straight lines. With any two points, we can immediately tell how much a line is increasing or decreasing, as seen in Figure 2.7.

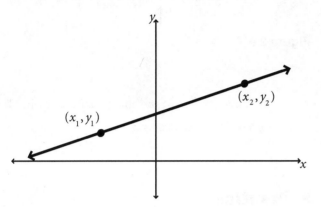

**Figure 2.7**

"How much a line is increasing or decreasing" is called the *slope* and is calculated by dividing "rise over run":

$$slope = \frac{rise}{run}$$

$$= \frac{y\text{-}change}{x\text{-}change}$$

$$= \frac{y_2 - y_1}{x_2 - x_1}$$

### Example
What is the slope of the line through points (2,7) and (−1,5)?

### Solution
$$slope = \frac{5 - 7}{-1 - 2} = \frac{-2}{-3} = \frac{2}{3}$$

## ▶ Practice

Find the slope between the following points.
**21.** (1,5) and (2,8)
**22.** (2,5) and (6,7)
**23.** (7,3) and (−2,3)
**24.** (−2,−4) and (−6,5)
**25.** (2,7) and (5,w)
**26.** (4,10) and (x,y)

## ▶ Point-Slope Formula

The most wonderful thing about straight lines is that their slopes are always the same. Thus, if a straight line has slope $m$ and goes through the point $(x_1, y_1)$, then any other point $(x,y)$ on the line will calculate the same slope:
$$\frac{y - y_1}{x - x_1} = m$$

By cross-multiplying, we get the *point-slope formula* for finding the equation of a straight line:
$$y - y_1 = m(x - x_1)$$
or equivalently
$$y = m(x - x_1) + y_1$$
Here, $y$ is a function of $x$, which could be written as
$$y(x) = m(x - x_1) + y_1.$$

### Example
Find the equation of the line with slope −2 through point (−1,8). Graph the line.

## Solution

$$y = -2(x - (-1)) + 8$$

$$y = -2x + 6$$

This form of the equation is called the *slope-intercept* form because $-2$ is the slope and 6 is where the line intercepts the $y$-axis (see Figure 2.8):

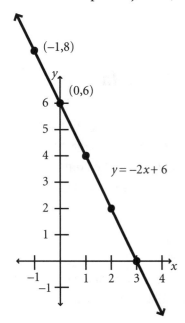

**Figure 2.8**

The slope of $-2 = \dfrac{-2}{1}$ means the $y$-value goes down 2 with every 1 increase in the $x$-value.

## Example

Find the equation of the straight line through $(2,6)$ and $(5,7)$. Graph the line.

## Solution

The slope is $\dfrac{7 - 6}{5 - 2} = \dfrac{1}{3}$, so the equation is $y = \dfrac{1}{3}(x - 2) + 6 = \dfrac{1}{3}x + \dfrac{16}{3}$ (see Figure 2.9).

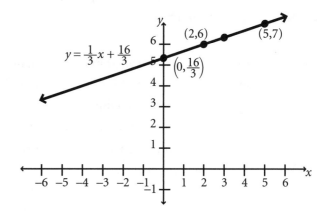

**Figure 2.9**

The slope of $\dfrac{1}{3}$ means the $y$-value goes up 1 when the $x$-value increases by 3.

## ▶ Practice

Find the equation of the straight line with the given information and then graph the line.

**27.** slope 2 through point $(1,-2)$

**28.** slope $-\dfrac{2}{3}$ through point $(6,1)$

**29.** through points $(5,3)$ and $(-1,-3)$

**30.** through points $(2,5)$ and $(6,5)$

# 3 ▶ Exponents and Logarithms

## ▶ Exponents

Because exponents form such an important part of calculus, we shall briefly review them. Generally, $a^n$ means "multiply the base $a$ as many times as the exponent $n$."

$$a^n = \underbrace{a \cdot a \cdot a \cdots a}_{n \text{ times}}$$

**Note:** The exponent formulas in this lesson all assume that $a$ is a positive number.

### Examples
Review the following examples by multiplying out.

$3^4 = 3 \cdot 3 \cdot 3 \cdot 3 = 81$

$2^5 = 2 \cdot 2 \cdot 2 \cdot 2 \cdot 2 = 32$

$5^1 = 5$

$10^6 = 1,000,000$

A number to the first power is just that number:
$a^1 = a$

When two numbers with the same base are multiplied, their exponents are added.

$$a^n \cdot a^m = \underbrace{(a \cdot a \cdot a \cdots a)}_{n \text{ times}} \cdot \underbrace{(a \cdot a \cdot a \cdots a)}_{m \text{ times}} = a^{n+m}$$

## Examples
Review and simplify the following.

$4^{10} \cdot 4^7 = 4^{17}$

$10^2 \cdot 10^5 = 10^7$

$5^3 \cdot 5 = 5^3 \cdot 5^1 = 5^4$

$7^2 \cdot 7^4 \cdot 7^3 = 7^9$

The rule about adding exponents has an interesting consequence. We know that $\sqrt{5} \cdot \sqrt{5} = 5$ because this is what "square root" means. Also, however, $5^{\frac{1}{2}} \cdot 5^{\frac{1}{2}} = 5^{\frac{1}{2}+\frac{1}{2}} = 5^1 = 5$. Because $\sqrt{5}$ and $5^{\frac{1}{2}}$ act exactly the same, they are equal: $\sqrt{5} = 5^{\frac{1}{2}}$. This works for square roots, cube roots, and so on:

$$a^{\frac{1}{2}} = \sqrt{a}, \quad a^{\frac{1}{3}} = \sqrt[3]{a}, \quad a^{\frac{1}{4}} = \sqrt[4]{a}, \ldots$$

## Examples
Simplify the following.

$9^{\frac{1}{2}} = \sqrt{9} = 3$

$63^{\frac{1}{3}} = \sqrt[3]{64} = 4$

When two numbers with the same base are divided, their exponents are subtracted.

$$\frac{a^n}{a^m} = a^{n-m}$$

## Examples
Work through the following simplifications.

$$\frac{3^5}{3^2} = \frac{3 \cdot 3 \cdot 3 \cdot 3 \cdot 3}{3 \cdot 3} = \frac{3 \cdot 3 \cdot 3 \cdot \cancel{3} \cdot \cancel{3}}{\cancel{3} \cdot \cancel{3}} = 3 \cdot 3 \cdot 3 = 3^3$$

$$\frac{11^{15}}{11^6} = 11^9$$

The rule about subtracting exponents has two interesting consequences. First, $\frac{5^4}{5^4} = 1$ because any nonzero number divided by itself is one. Also, $\frac{5^4}{5^4} = 5^{4-4} = 5^0$. Thus, $5^0 = 1$. In general:

$$a^0 = 1$$

Simplify the following.

$3^0 = 1$

$200^0 = 1$

The second consequence follows from:

$$\frac{2^3}{2^7} = \frac{\cancel{2} \cdot \cancel{2} \cdot \cancel{2}}{2 \cdot 2 \cdot 2 \cdot 2 \cdot \cancel{2} \cdot \cancel{2} \cdot \cancel{2}} = \frac{1}{2 \cdot 2 \cdot 2 \cdot 2} = \frac{1}{2^4} \text{ while}$$

also $\frac{2^3}{2^7} = 2^{3-7} = 2^{-4}$. Thus, $2^{-4} = \frac{1}{2^4}$. In general:

$$a^{-n} = \frac{1}{a^n}$$

## Examples
Work through the following simplifications.

$$3^{-2} = \frac{1}{3^2} = \frac{1}{9}$$

$$4^{-1} = \frac{1}{4^1} = \frac{1}{4}$$

$$5^{-\frac{1}{2}} = \frac{1}{5^{\frac{1}{2}}} = \frac{1}{\sqrt{5}}$$

# ▶ Practice

Simplify the following.

**1.** $2^3 \cdot 2^2$

**2.** $4 \cdot 4^2$

**3.** $\dfrac{10^7}{10^3}$

**4.** $\dfrac{6^3}{6^5}$

**5.** $6^0$

**6.** $3^8 \cdot 3 \cdot 3^{-5}$

**7.** $9^1$

**8.** $25^{\frac{1}{2}}$

**9.** $5^{-1}$

**10.** $8^{\frac{1}{3}}$

**11.** $2^{-3}$

**12.** $8^{\frac{2}{3}}$

**13.** $\dfrac{1}{5^{-1}}$

**14.** $10^{-5}$

**15.** $\dfrac{1}{8^{-2}}$

**16.** $\dfrac{1}{16^{-\frac{1}{2}}}$

# ▶ Exponential Functions

We can form an *exponential function* by leaving the base fixed and varying the exponent.

## Example

The function $f(x) = 2^x$ has the graph shown in Figure 3.1. Note that $2^x$ is quite different from $x^2$. For example, when $x = 10$, the value of $2^x$ is $2^{10} = 2 \cdot 2 \cdot 2 \cdot 2 \cdot 2 \cdot 2 \cdot 2 \cdot 2 \cdot 2 \cdot 2 = 1{,}024$, while the value of $x^2$ is $10^2 = 10 \cdot 10 = 100$.

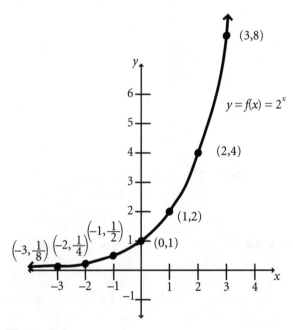

**Figure 3.1**

## Example

The function $g(x) = 3^x$ has the graph shown in Figure 3.2. For reasons that will become clear later, a very nice base to use is the number $e = 2.71828\ldots$, which, just like $\pi = 3.14159\ldots$, can never be written out completely.

The exponential function takes $x$ to $e^x$ and the natural logarithm takes it right back to $x$, so $\ln(e^x) = x$.

Similarly, $e^{\ln(x)} = x$.

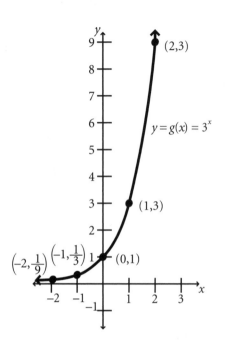

**Figure 3.2**

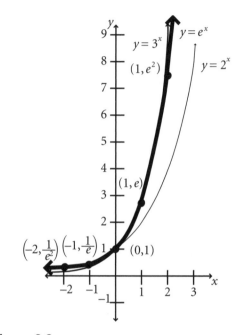

**Figure 3.3**

Because $2 < e < 3$, the graph of $y = e^x$ fits between $y = 2^x$ and $y = 3^x$ (see Figure 3.3).

Other than the strange base, everything about $e^x$ is normal.

$$e^0 = 1$$

$$e^n \cdot e^m = e^{n+m}$$

$$e^1 = e$$

$$\frac{e^n}{e^m} = e^{n-m}$$

Another useful function is the opposite of $e^x$, known as the *natural logarithm* $\ln(x)$. Just as subtracting undoes adding, dividing undoes multiplying, and taking a square root undoes squaring, the natural logarithm undoes $e^x$.

If $y = e^x$, then $\ln(y) = \ln(e^x)$, so $\ln(y) = x$.

The graph of $y = \ln(x)$ comes from flipping the graph of $y = e^x$ across the line $y = x$, as depicted in Figure 3.4.

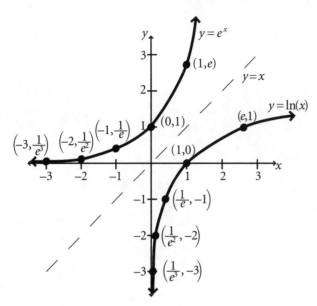

**Figure 3.4**

The laws of $\ln(x)$ are rather unusual.

$$\ln(a) + \ln(b) = \ln(a \cdot b)$$

$$\ln(a) - \ln(b) = \ln\left(\frac{a}{b}\right)$$

$$\ln(a^n) = n \cdot \ln(a)$$

The last of the three preceding laws is useful for turning an exponent into a matter of multiplication.

## Example
Solve for $x$ when $10^x = 7$.

## Solution
Take the natural logarithm of both sides.

$$\ln(10^x) = \ln(7)$$

Use $\ln(a^n) = n \cdot \ln(a)$.

$$x \cdot \ln(10) = \ln(7)$$

Divide both sides by $\ln(10)$.

$$x = \frac{\ln(7)}{\ln(10)}$$

A calculator can be used to find a decimal approximation: $\frac{\ln(7)}{\ln(10)} \approx 0.84509$, if desired.

## Example
Simplify $\ln(25) + \ln(4) - \ln(2)$.

## Solution
Use $\ln(a) + \ln(b) = \ln(a \cdot b)$.

$$\ln(25 \cdot 4) - \ln(2)$$

Use $\ln(a) - \ln(b) = \ln\left(\frac{a}{b}\right)$.

$$\ln\left(\frac{25 \cdot 4}{2}\right) = \ln(50)$$

## ▶ Practice

Simplify the following.

**17.** $e^3 \cdot e^8$

**18.** $\dfrac{e^{12}}{e^5}$

**19.** $e^0$

**20.** $\ln(e^2)$

**21.** $e^{\ln(5)}$

**22.** $\ln(7) + \ln(2)$

**23.** $\ln(24) - \ln(6)$

**24.** Solve for $x$ when $2^x = 10$.

**25.** Solve for $x$ when $8^x = 11$.

**26.** Solve for $x$ when $3^x \cdot 3^5 = 100$.

# 4 ▶ Trigonometry

Some very interesting and important functions are formed by dividing the length of one side of a right triangle by the length of another side. These functions are called *trigonometric* because they come from the geometry of a triangle. The domain consists of the measures $x$ of angles. Let $H$ represent the length of the *hypotenuse*, $A$ represent the length of the side *adjacent* to the angle $x$, and the letter $O$ represent the length of the side *opposite* (away) from the angle $x$. A right triangle with angle $x$ is depicted in Figure 4.1.

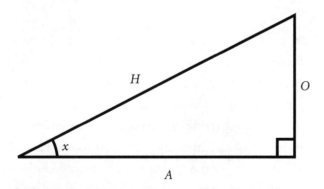

**Figure 4.1**

Some people remember the first three trigonometric functions by saying "Oliver Had A Heap Of Apples" to remember the $\frac{O}{H}$, $\frac{A}{H}$, and $\frac{O}{A}$ of sin(x), cos(x), and tan(x). Others say SOA CAH TOA to remember $\sin(x) = \frac{O}{A}$, $\cos(x) = \frac{A}{H}$, and $\tan(x) = \frac{O}{A}$.

---

The six trigonometric functions, *sine* (abbreviated sin), *cosine* (cos), *tangent* (tan), *secant* (sec), *cosecant* (csc), and *cotangent* (cot), are defined for each angle x by dividing the following sides:

$$\sin(x) = \frac{O}{H}$$

$$\cos(x) = \frac{A}{H}$$

$$\tan(x) = \frac{O}{A}$$

$$\sec(x) = \frac{H}{A}$$

$$\csc(x) = \frac{H}{O}$$

$$\cot(x) = \frac{A}{O}$$

The first thing to notice is that all of the functions can be obtained from just sin(x) and cos(x) using the following *trigonometric identities.*

$$\tan(x) = \frac{O}{A} = \frac{\frac{O}{H}}{\frac{A}{H}} = \frac{\sin(x)}{\cos(x)}$$

$$\sec(x) = \frac{H}{A} = \frac{1}{\frac{A}{H}} = \frac{1}{\cos(x)}$$

$$\csc(x) = \frac{H}{O} = \frac{1}{\frac{O}{H}} = \frac{1}{\sin(x)}$$

$$\cot(x) = \frac{A}{O} = \frac{\frac{A}{H}}{\frac{O}{H}} = \frac{\cos(x)}{\sin(x)}$$

Thus, all of the trigonometric functions can be evaluated for an angle x if the sin(x) and cos(x) are known.

The next thing to notice is that the Pythagorean theorem, which, stated in terms of the sides O, A, and H, is $O^2 + A^2 = H^2$. And, if we divide everything by $H^2$, we get the following:

$$\frac{O^2}{H^2} + \frac{A^2}{H^2} = \frac{H^2}{H^2}$$

$$\left(\frac{O}{H}\right)^2 + \left(\frac{A}{H}\right)^2 = 1$$

Thus, $(\sin(x))^2 + (\cos(x))^2 = 1$. To save on parentheses, we often write this as $\sin^2(x) + \cos^2(x) = 1$. Because no particular value of x was used in the calculations, this is true for every value of x.

Drawing triangles and measuring their sides is an impractical and inaccurate method to calculate the values of trigonometric functions. Most people use calculators instead. Although, when using a calculator, it is very important to make sure that it is set to the same format for measuring angles that you are already using: that is, *degrees* or *radians.*

## Conversion Hint

To convert from degrees to radians, multiply by $\dfrac{2\pi}{360} = \dfrac{\pi}{180}$.

To convert from radians to degrees, multiply by $\dfrac{360}{2\pi} = \dfrac{180}{\pi}$.

There are 360 degrees in a circle, possibly because ancient peoples thought that there were 360 days in a year. As the earth went around the sun, the position of the sun against the background stars moved one *degree* every day. The $2\pi$ radians in a circle correspond to the distance around a circle of radius 1. Because radians already correspond to a distance, there is no need for conversions when calculating with radians. Mathematicians thus use radians almost exclusively.

- To convert from degrees to radians, multiply by
  $$\frac{2\pi}{360} = \frac{\pi}{180}.$$

- To convert from radians to degrees, multiply by
  $$\frac{360}{2\pi} = \frac{180}{\pi}.$$

### Example
Convert $45°$ into radians.

### Solution
$$45° = 45 \cdot \frac{\pi}{180} \text{ radians} = \frac{\pi}{4} \text{ radians}$$

### Example
Convert $\dfrac{2\pi}{3}$ radians into degrees.

### Solution
$$\frac{2\pi}{3} \text{ radians} = \frac{2\pi}{3} \cdot \frac{180°}{\pi} = 120°$$

## ▶ Practice

Convert the following into radians.

**1.** $30°$

**2.** $180°$

**3.** $270°$

**4.** $300°$

**5.** $135°$

Convert the following into degrees.

**6.** $\dfrac{\pi}{3}$

**7.** $\dfrac{\pi}{2}$

**8.** $2\pi$

**9.** $\dfrac{\pi}{10}$

**10.** $\dfrac{4\pi}{2}$

## ▶ Trigonometric Values of Nice Angles

There are a few nice angles for which the trigonometric functions can be easily calculated. If $x = \dfrac{\pi}{4} = 45°$, then the two legs of the triangle are equal. If the hypotenuse is $H = 1$, then we have what you can see in Figure 4.2.

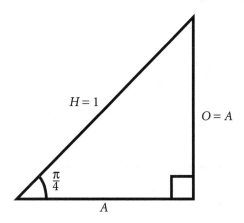

**Figure 4.2**

By the Pythagorean theorem, $A^2 + A^2 = 1$, so $2A^2 = 1$ and $A^2 = \frac{1}{2}$. This means that $O = A =$ $\frac{1}{\sqrt{2}} = \frac{1}{\sqrt{2}}$. If we rationalize the denominator, we get $\frac{1}{\sqrt{2}} = \frac{1}{\sqrt{2}} \cdot \frac{\sqrt{2}}{\sqrt{2}} = \frac{\sqrt{2}}{2}$. Thus:

$$\sin\left(\frac{\pi}{4}\right) = \frac{O}{H} = \frac{\frac{\sqrt{2}}{2}}{1} = \frac{\sqrt{2}}{2}$$

$$\cos\left(\frac{\pi}{4}\right) = \frac{A}{H} = \frac{\frac{\sqrt{2}}{2}}{1} = \frac{\sqrt{2}}{2}$$

Another nice angle is $x = 60° = \frac{\pi}{3}$, because it is found in equilateral triangles such as those seen in Figure 4.3. This triangle can be cut in half to form the triangle shown in Figure 4.4.

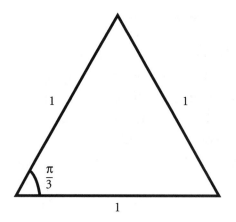

**Figure 4.3**

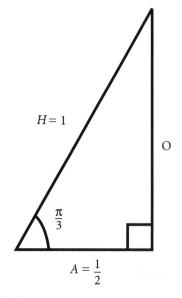

**Figure 4.4**

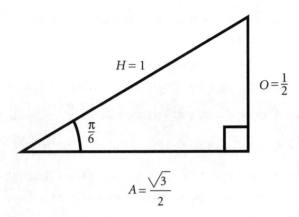

**Figure 4.5**

By the Pythagorean theorem, $\left(\dfrac{1}{2}\right)^2 + O^2 = 1^2$,

so $O^2 = 1 - \dfrac{1}{4} = \dfrac{3}{4}$. Thus, $O = \sqrt{\dfrac{3}{4}} = \dfrac{\sqrt{3}}{2}$. This

means that:

$$\sin\left(\frac{\pi}{3}\right) = \frac{O}{H} = \frac{\frac{\sqrt{3}}{2}}{1} = \frac{\sqrt{3}}{2}$$

$$\cos\left(\frac{\pi}{3}\right) = \frac{A}{H} = \frac{\frac{1}{2}}{1} = \frac{1}{2}$$

We can flip that last triangle around to calculate the trigonometric functions for the other angle $x = 30° = \dfrac{\pi}{6}$ (see Figure 4.5).

$$\sin\left(\frac{\pi}{6}\right) = \frac{O}{H} = \frac{\frac{1}{2}}{1} = \frac{1}{2}$$

$$\cos\left(\frac{\pi}{6}\right) = \frac{A}{H} = \frac{\frac{\sqrt{3}}{2}}{1} = \frac{\sqrt{3}}{2}$$

### Example

Use the trigonometric identities to find $\sec\left(\dfrac{\pi}{4}\right)$.

### Solution

Use the trigonometric identity for sec.

$$\sec(x) = \frac{1}{\cos(x)}$$

Use $x = \dfrac{\pi}{4}$.

$$\sec\left(\frac{\pi}{4}\right) = \frac{1}{\cos(\frac{\pi}{4})}$$

Use $\cos\left(\dfrac{\pi}{4}\right) = \dfrac{\sqrt{2}}{2}$.

$$\sec\left(\frac{\pi}{4}\right) = \frac{1}{\frac{\sqrt{2}}{2}}$$

Simplify.

$$\sec\left(\frac{\pi}{4}\right) = \frac{2}{\sqrt{2}} = \sqrt{2}$$

### ▶ Practice

Use the trigonometric identities to evaluate the following.

**11.** $\tan\left(\dfrac{\pi}{4}\right)$

**12.** $\tan\left(\dfrac{\pi}{3}\right)$

**13.** $\csc\left(\dfrac{\pi}{6}\right)$

**14.** $\sec\left(\dfrac{\pi}{3}\right)$

**15.** $\cot\left(\dfrac{\pi}{3}\right)$

**16.** $\cot\left(\dfrac{\pi}{6}\right)$

**17.** $\sec\left(\dfrac{\pi}{6}\right)$

**18.** $\csc\left(\dfrac{\pi}{4}\right)$

## ▶ Trigonometric Values for Angles Greater Than $90° = \dfrac{\pi}{2}$

Notice that when the hypotenuse has length 1, the sine of the angle is the height of the triangle and the cosine is the width. Equivalently, the sine is the $y$-value of the point where a ray of the given angle intersects with the circle of radius 1. Similarly, the cosine is the $x$-value.

### Example

For example, when $x = \dfrac{\pi}{6} = 30°$, we have the picture shown in Figure 4.6.

$$\sin\left(\frac{\pi}{6}\right) = \frac{1}{2}$$

$$\cos\left(\frac{\pi}{6}\right) = \frac{\sqrt{3}}{2}$$

The circle of radius 1 around the origin is called the *unit circle*. Note in Figure 4.7 that the angle of measure 0 runs straight to the right along the positive $x$-axis, and every other angle is measured counter-clockwise up from this.

This can be used to find the trigonometric values of nice angles greater than $90° = \dfrac{\pi}{2}$. The trick is to use either a $30°, 60°, 90°$ triangle (a $\dfrac{\pi}{6}, \dfrac{\pi}{3}, \dfrac{\pi}{2}$ triangle) or else a $45°, 45°, 90°$ triangle (a $\dfrac{\pi}{4}, \dfrac{\pi}{4}, \dfrac{\pi}{2}$ triangle) to find the $y$-value(sine) and $x$-value (cosine) of the appropriate point on the unit circle. As before, calculating the trigonometric values for non-nice angles requires the help of either much more mathematics or the use of a calculator.

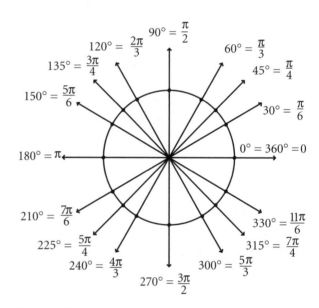

**Figure 4.7**

### Example

Find the sine and cosine of $120° = \dfrac{2\pi}{3}$.

### Solution

For this angle, we use a $\dfrac{\pi}{6}, \dfrac{\pi}{3}, \dfrac{\pi}{2}$ triangle, as shown in Figure 4.8 to find the $x$- and $y$-values. The $y$-value of the point where the ray of angle $\dfrac{2\pi}{3}$ hits

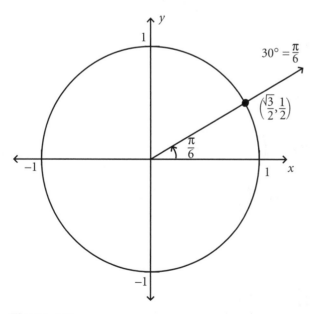

**Figure 4.6**

the unit circle is $y = \dfrac{\sqrt{3}}{2}$. Thus, $\sin\left(\dfrac{2\pi}{3}\right) = \dfrac{\sqrt{3}}{2}$.

The $x$-value is negative, $x = -\dfrac{1}{2}$, so $\cos\left(\dfrac{2\pi}{3}\right) = -\dfrac{1}{2}$.

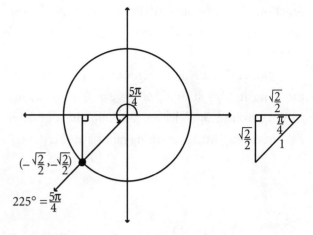

Figure 4.9

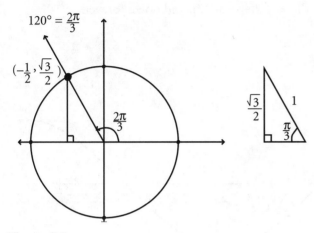

**Figure 4.8**

## Example

Find the sine and cosine of $\dfrac{5\pi}{4} = 225°$.

## Solution

Because $225°$ is a multiple of $45°$, we use a $45°, 45°, 90°$ triangle to find the $x$- and $y$-values. As seen in Figure 4.9, both of the coordinates are negative, so $\sin\left(\dfrac{5\pi}{4}\right) = -\dfrac{\sqrt{2}}{2}$ and $\cos\left(\dfrac{5\pi}{4}\right) = -\dfrac{\sqrt{2}}{2}$ are both negative.

## Example

Find all of the trigonometric values for $90° = \dfrac{\pi}{2}$.

## Solution

Even though there isn't a triangle here, there is still a point on the unit circle. See Figure 4.10. We conclude that $\cos\left(\dfrac{\pi}{2}\right) = 0$ and $\sin\left(\dfrac{\pi}{2}\right) = 1$ from the $x$- and $y$-values of the point. Using the trigonometric identities, we can calculate that $\csc\left(\dfrac{\pi}{2}\right) = \dfrac{1}{\sin(\frac{\pi}{2})} = 1$ and $\cot\left(\dfrac{\pi}{2}\right) = \dfrac{\cos(\frac{\pi}{2})}{\sin(\frac{\pi}{2})} = \dfrac{0}{1} = 0$. The tangent and secant

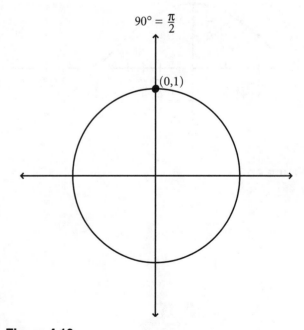

**Figure 4.10**

functions, however, involve division by 0 and thus are left undefined. The angle $x = \frac{\pi}{2}$ is not in the domain of tan and sec.

Notice that all of the trigonometric functions are the same at $0° = 0$ and $360° = 2\pi$. This is because turning around $360°$ leaves you facing in your original direction. Thus, everything repeats at this point.

Using the table along with the fact that everything repeats, we can sketch the graphs of $\sin(x)$ and $\cos(x)$. See Figures 4.11 and 4.12.

The functions sine and cosine are classic examples of repeating, or *oscillating*, functions because of the way they wave up and down forever.

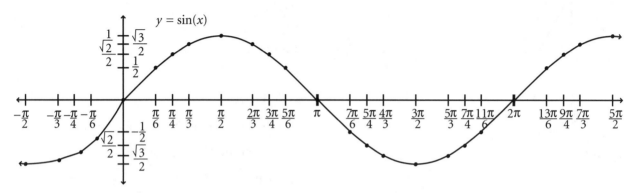

**Figure 4.11**

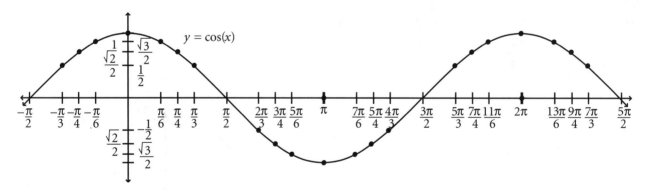

**Figure 4.12**

# ▶ Practice

Use the unit circle and the trigonometric identities to complete the following table. Find the answers to questions 19 through 36.

| | $\sin(x)$ | $\cos(x)$ | $\tan(x)$ | $\sec(x)$ | $\csc(x)$ | $\cot(x)$ |
|---|---|---|---|---|---|---|
| $0° = 0$ | $0$ | $1$ | $0$ | $1$ | undef. | undef. |
| $30° = \frac{\pi}{6}$ | $\frac{1}{2}$ | $\frac{\sqrt{3}}{2}$ | *19* | $\frac{2\sqrt{3}}{3}$ | $2$ | $\sqrt{3}$ |
| $45° = \frac{\pi}{4}$ | $\frac{\sqrt{2}}{2}$ | $\frac{\sqrt{2}}{2}$ | $1$ | $\sqrt{2}$ | $\sqrt{2}$ | $1$ |
| $60° = \frac{\pi}{3}$ | $\frac{\sqrt{3}}{2}$ | $\frac{1}{2}$ | $\sqrt{3}$ | $2$ | $\frac{2\sqrt{3}}{3}$ | $\frac{\sqrt{3}}{3}$ |
| $90° = \frac{\pi}{2}$ | $1$ | $0$ | undef. | undef. | $1$ | $0$ |
| $120° = \frac{2\pi}{3}$ | $\frac{\sqrt{3}}{2}$ | $-\frac{1}{2}$ | $-\sqrt{3}$ | $-2$ | *20* | $-\frac{\sqrt{3}}{3}$ |
| $135° = \frac{3\pi}{4}$ | *21* | *22* | *23* | *24* | *25* | *26* |
| $150° = \frac{5\pi}{6}$ | $\frac{1}{2}$ | $-\frac{\sqrt{3}}{2}$ | $-\frac{\sqrt{3}}{3}$ | $-\frac{2\sqrt{3}}{3}$ | $2$ | $-\sqrt{3}$ |
| $180° = \pi$ | *27* | $-1$ | $0$ | $-1$ | *28* | undef. |
| $210° = \frac{7\pi}{6}$ | $-\frac{1}{2}$ | $-\frac{\sqrt{3}}{2}$ | $\frac{\sqrt{3}}{3}$ | $-\frac{2\sqrt{3}}{3}$ | $-2$ | *29* |
| $225° = \frac{5\pi}{4}$ | $-\frac{\sqrt{2}}{2}$ | $-\frac{\sqrt{2}}{2}$ | $1$ | $-\sqrt{2}$ | $-\sqrt{2}$ | $1$ |
| $240° = \frac{4\pi}{3}$ | *30* | *31* | *32* | *33* | *34* | *35* |
| $270° = \frac{3\pi}{2}$ | $-1$ | $0$ | undef. | undef. | $-1$ | $0$ |
| $300° = \frac{5\pi}{3}$ | $-\frac{\sqrt{3}}{2}$ | $\frac{1}{2}$ | $-\sqrt{3}$ | $2$ | $-\frac{2\sqrt{3}}{3}$ | $-\frac{\sqrt{3}}{3}$ |
| $315° = \frac{7\pi}{4}$ | $-\frac{\sqrt{2}}{2}$ | $\frac{\sqrt{2}}{2}$ | $-1$ | *36* | $-\sqrt{2}$ | $-1$ |
| $360° = 2\pi$ | $0$ | $1$ | $0$ | $1$ | undef. | undef. |

**Note:** The numbers appearing in bold with asterisks are questions 19 through 36.

**19.** Find the value that goes in the position in the table where you see *19*.

**20.** Find the value that goes in the position in the table where you see *20*.

**21.** Find the value that goes in the position in the table where you see *21*.

**22.** Find the value that goes in the position in the table where you see *22*.

**23.** Find the value that goes in the position in the table where you see *23*.

**24.** Find the value that goes in the position in the table where you see *24*.

**25.** Find the value that goes in the position in the table where you see *25*.

**26.** Find the value that goes in the position in the table where you see *26*.

**27.** Find the value that goes in the position in the table where you see *27*.

**28.** Find the value that goes in the position in the table where you see *28*.

**29.** Find the value that goes in the position in the table where you see *29*.

**30.** Find the value that goes in the position in the table where you see *30*.

**31.** Find the value that goes in the position in the table where you see *31*.

**32.** Find the value that goes in the position in the table where you see *32*.

**33.** Find the value that goes in the position in the table where you see *33*.

**34.** Find the value that goes in the position in the table where you see *34*.

**35.** Find the value that goes in the position in the table where you see *35*.

**36.** Find the value that goes in the position in the table where you see *36*.

# 5 ▶ Limits

**M**athematicians, just like children, like to see what happens when we push limits. We are told not to divide by zero, so the temptation overwhelms us to see what happens when we divide by *almost* zero. The process of using *almost* numbers underlies the concept of a limit.

Limits can be most easily seen graphically. For example, look at the graph of $y = f(x)$ in Figure 5.1.

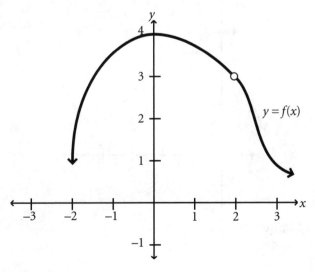

**Figure 5.1**

The domain of *f* is $x \neq 2$. We can't plug $x = 2$ into *f*. However, the hole is in a clear place, at $(2,3)$. How do we know the hole has a *y*-value of 3? Well, the points on the curve with *x*-values *near* $x = 2$ have *y*-values *close* to $y = 3$. The closer we get to $x = 2$, the closer the *y*-values of the points come to $y = 3$.

The mathematical shorthand for this is $\lim_{x \to 2} f(x) = 3$, which is pronounced "the limit as *x* approaches 2 of $f(x)$ is 3."

We don't have to just approach discontinuities, though. For example, $\lim_{x \to 0} f(x) = 4$. Note that this is a statement about the points *near* $x = 0$ having $f(x)$ *near* 4. The exact point at $(0,4)$ isn't used in evaluating the limit.

We can also approach points from either the left or from the right. For example, take Figure 5.2 to be the graph of $y = g(x)$.

Here, $\lim_{x \to 1^-} g(x) = 4$ and $\lim_{x \to 1^+} g(x) = 2$. The little minus in $\lim_{x \to 1^+}$ means that we approach $x = 1$ using numbers less than (to the left) of $x = 1$. As we approach $x = 1$ from the left-hand side, we slide up the graph through *y*-values that approach 4. Similarly,

the plus in $\lim_{x \to 1^+}$ means "approach from the right." From the right, the height of the graph slides down to $y = 2$ as *x* approaches 1.

In this example, $\lim_{x \to 1} g(x)$ does not exist because there is no single *y*-value to which all of the points near $x = 1$ get close. Some are close to 4, and others are close to 2. Because there is no agreement, there is no limit.

As another example, let Figure 5.3 be the graph of $y = h(x)$. Here, $\lim_{x \to 3^-} h(x) = 2$ because sliding up to $x = 3$ from the left has us pass through points with *y*-values near 2. Similarly, $\lim_{x \to 3^+} h(x) = 2$. Because there *is* agreement from the left and right, we have the general limit, $\lim_{x \to 3} h(x) = 2$. Once again, notice that what happens at exactly $x = 3$ is irrelevant. Here $h(3) = 5$, but the resulting point at $(3,5)$ has no bearing on the limit of points *approaching* $x = 3$.

Vertical asymptotes correspond with infinite limits. For example, take the graph in Figure 5.4 of $y = k(x)$.

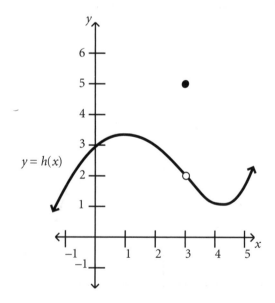

**Figure 5.3**

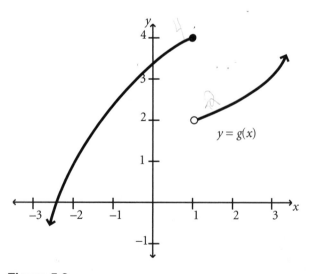

**Figure 5.2**

at $x = a$) and the limit from either side goes up to the same value, then the function flows continuously through that point.

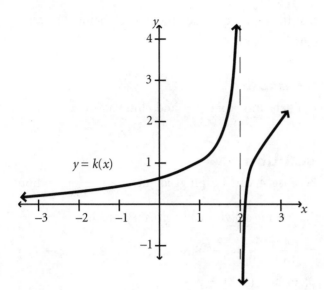

$y = k(x)$

**Figure 5.4**

Here, $\lim\limits_{x \to 2^-} k(x) = \infty$, $\lim\limits_{x \to 2^+} k(x) = -\infty$, and $\lim\limits_{x \to 2} k(x)$ does not exist.

All of these examples involve discontinuities. We can rule them out in the following manner. If $\lim\limits_{x \to a^-} f(x) = f(a) = \lim\limits_{x \to a^+} f(x)$, then $f$ is continuous at $x = a$. In other words, if $f(a)$ exists (there is a point

## ▶ Practice

Use Figure 5.5 to evaluate the following.

**1.** $\lim\limits_{x \to -1^-} f(x)$

**2.** $\lim\limits_{x \to -1^+} f(x)$

**3.** $\lim\limits_{x \to -1} f(x)$

**4.** $f(-1)$

**5.** Is $f$ continuous at $x = -1$?

**6.** $\lim\limits_{x \to 3^-} f(x)$

**7.** $\lim\limits_{x \to 3^+} f(x)$

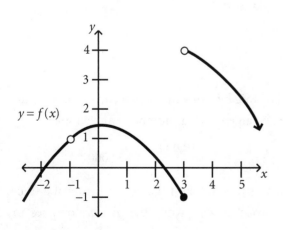

$y = f(x)$

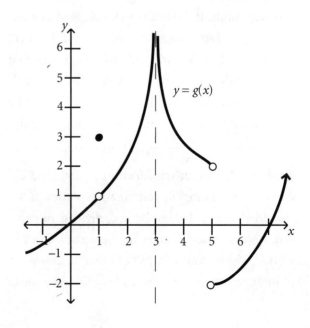

$y = g(x)$

**Figure 5.5**

**8.** $\lim\limits_{x \to 3} f(x)$

**9.** $f(3)$

**10.** Is $f$ continuous at $x = 1$?

**11.** $\lim\limits_{x \to 1} g(x)$

**12.** $g(1)$

**13.** $\lim\limits_{x \to 3^-} g(x)$

**14.** $\lim\limits_{x \to 3} g(x)$

**15.** $\lim\limits_{x \to 5^-} g(x)$

**16.** $\lim\limits_{x \to 5^+} g(x)$

## ► Evaluating Limits Algebraically

It is not necessary to have the graph of a function to evaluate its limits. If the limit can be plugged in without dividing by zero, that is how the limit is calculated.

Technically, this works only with functions that are *polynomials* (formed by a variable added and multiplied with constants) like $4x^5 - 10x^3 - 7$, roots like $\sqrt{x}$, *rational functions* (formed by dividing two polynomials) like $\dfrac{3x - 5}{2x^3 + x^2 + 1}$, and *transcendental functions* like the trigonometric functions, $\ln(x)$, and $e^x$. Because this works for any combination of these functions added, subtracted, multiplied, divided, or composed, it works also for every function considered in this book. The reason is that all of these functions are continuous on their domains, and continuity ensures

that the limits approach the points obtained by plugging in values.

### Example
Evaluate $\lim\limits_{x \to 4} \dfrac{x + 5}{x^2 + 10x}$ and $\lim\limits_{x \to -2} 3x^2 + x - 7$.

### Solution
Because 4 can be plugged into $\dfrac{x + 5}{x^2 + 10x}$ without there being a division by zero, the limit $\lim\limits_{x \to 4} \dfrac{x + 5}{x^2 + 10x}$ $= \dfrac{4 + 5}{16 + 40} = \dfrac{9}{56}$. Similarly, $\lim\limits_{x \to -2} 3x^2 + x - 7 = 3(-2)^2 + (-2) - 7 = 3$.

## ► Practice

Evaluate the following limits.

**17.** $\lim\limits_{x \to 1} 10x^3 + 4x^2 - 5x + 7$

**18.** $\lim\limits_{x \to 2} \dfrac{x^3 - 4}{10x + 3}$

**19.** $\lim\limits_{x \to 3} \dfrac{x - 3}{x^2 + x}$

**20.** $\lim\limits_{x \to \frac{\pi}{6}} \dfrac{\sin(x)}{x}$

**21.** $\lim\limits_{a \to 0} 2x + a + 1$

**22.** $\lim\limits_{a \to 0} 3x^2 + 3xa + a^2$

Dividing by a tiny number is equivalent to multiplying by an enormous number. For example:

$$5 \div \dfrac{1}{10,000} = 5 \cdot \dfrac{10,000}{1} = 50,000$$

It is for this reason that if the denominator of a fraction approaches zero while the numerator goes to something nonzero, the result is an infinite limit.

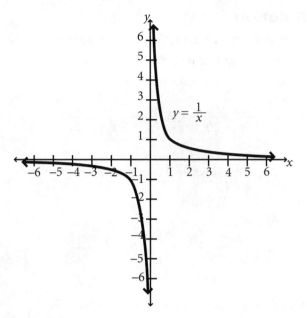

**Figure 5.6**

The classic example is $f(x) = \dfrac{1}{x}$ (graphed in Figure 5.6). This has the following limits at zero: $\lim\limits_{x \to 0^-} \dfrac{1}{x} = -\infty$, $\lim\limits_{x \to 0^+} \dfrac{1}{x} = \infty$, and $\lim\limits_{x \to 0} \dfrac{1}{x}$ does not exist.

Because the denominator goes to zero while the numerator stays one in all of these cases, there is a vertical asymptote at $x = 0$. The function therefore approaches either positive or negative infinity from either side. When $x$ is less than zero, as it always is when $x \to 0^-$, the function $\dfrac{1}{x}$ is also negative. Thus, $\lim\limits_{x \to 0^-} \dfrac{1}{x} = -\infty$. Similarly, as $x \to 0^+$, $\dfrac{1}{x}$ is always positive, so $\lim\limits_{x \to 0^+} \dfrac{1}{x} = \infty$. Finally, because the limit from the two sides are different, the undirected limit $\lim\limits_{x \to 0} \dfrac{1}{x}$ does not exist.

### Example

Evaluate $\lim\limits_{x \to 2} \dfrac{x + 3}{(x - 2)(x - 4)}$.

### Solution

The numerator approaches 5 while the denominator approaches 0. Therefore, this limit from the right is either $\infty$ or $-\infty$. What we need to figure out is whether the function is positive or negative at $x$-values just slightly larger than 2. We do this by looking at each factor individually.

As $x \to 2^+$, the factor $(x + 3) \to 5^+$ (a positive number), $(x - 2) \to 0^+$ (a positive number), and $(x - 4) \to -2^+$ (a negative number). Because the function $\dfrac{x + 3}{(x - 2)(x - 4)}$ is made of two positive parts and one negative part, the result will be negative. Thus, $\lim\limits_{x \to 2^+} \dfrac{x + 3}{(x - 2)(x - 4)} = -\infty$.

There are other, perhaps easier, ways to evaluate such limits. One is to plug into the function a representative number. In the previous example, for instance, when $x = 2.01$, the function is $\dfrac{(2.01) + 3}{((2.01) - 2)((2.01) - 4)} \approx -251$. Because this is negative, the limit is $-\infty$. Another method will be covered in Lesson 13.

### Example

Evaluate $\lim\limits_{x \to -3^-} \dfrac{(x + 1)(2 - x)}{(x + 3)(x + 5)}$.

### Solution

Here, the numerator approaches $-10$, which isn't zero, while the denominator approaches zero, so the limit is either $\infty$ or $-\infty$. While $x \to -3^-$, the factors:

$(x + 1) \to -2^-$ (negative)

$(2 - x) \to 5^+$ (positive)

$(x + 3) \to 0^-$ (negative)

$(x + 5) \to 2^-$ (positive)

The combination of two negative factors and two positive factors is positive, thus:

$$\lim_{x \to -3^-} \frac{(x + 1)(2 - x)}{(x + 3)(x + 5)} = \infty$$

## ▶ Practice

Evaluate the following limits.

**23.** $\displaystyle\lim_{x \to 1^-} \frac{1}{x - 1}$

**24.** $\displaystyle\lim_{x \to 4^+} \frac{x + 5}{x - 4}$

**25.** $\displaystyle\lim_{x \to 3^+} \frac{x - 2}{x + 3}$

**26.** $\displaystyle\lim_{x \to 3^+} \frac{(x + 2)(x - 5)}{(x + 6)(x - 3)}$

**27.** $\displaystyle\lim_{x \to 2} \frac{(x + 5)(x - 5)}{(x - 3)(x + 4)}$

**28.** $\displaystyle\lim_{x \to -5} \frac{x - 2}{(x + 5)^2}$

When both the numerator *and* the denominator go to zero, then there are two common tricks for simplifying the limit. The first is to factor. The second is to rationalize. The following example utilizes the first trick—factoring.

### Example

Evaluate $\displaystyle\lim_{x \to 4} \frac{x^2 - 2x - 8}{x^2 + x - 20}$.

### Solution

Here, both the numerator and denominator go to zero, so we aren't guaranteed an infinite limit. First, factor the numerator and denominator.

$$\lim_{x \to 4} \frac{x^2 - 2x - 8}{x^2 + x - 20} = \lim_{x \to 4} \frac{(x - 4)(x + 2)}{(x - 4)(x + 5)}$$

Because $x \neq 4$ as $x \to 4$, we can cancel $\dfrac{x - 4}{x - 4} = 1$.

$$\lim_{x \to 4} \frac{x^2 - 2x - 8}{x^2 + x - 20} =$$

$$\lim_{x \to 4} \frac{(x - 4)(x + 2)}{(x - 4)(x + 5)} = \lim_{x \to 4} \frac{(x + 2)}{(x + 5)}$$

Now we can plug in without dividing by zero.

$$\lim_{x \to 4} \frac{x^2 - 2x - 8}{x^2 + x - 20} = \lim_{x \to 4} \frac{(x + 2)}{(x + 5)} = \frac{6}{9} = \frac{2}{3}$$

The following example utilizes the trick of rationalizing.

### Example

Evaluate $\displaystyle\lim_{x \to 9} \frac{\sqrt{x} - 3}{x - 9}$.

### Solution

Because both numerator and denominator go to zero, a trick is necessary. First, multiply the top and bottom by the part with the square root, but with the opposite sign between them.

$$\lim_{x \to 9} \frac{\sqrt{x} - 3}{x - 9} = \lim_{x \to 9} \left( \frac{\sqrt{x} - 3}{x - 9} \right) \cdot \left( \frac{\sqrt{x} + 3}{\sqrt{x} + 3} \right)$$

Simplify.

$$\lim_{x \to 9} \frac{\sqrt{x} - 3}{x - 9} = \lim_{x \to 9} \frac{x + 3\sqrt{x} - 3\sqrt{x} - 9}{(x - 9)(\sqrt{x} + 3)}$$

Eliminate $\dfrac{x - 9}{x - 9} = 1$.

$$\lim_{x \to 9} \frac{\sqrt{x} - 3}{x - 9} = \lim_{x \to 9} \frac{\cancel{(x - 9)}}{\cancel{(x - 9)}(\sqrt{x} + 3)}$$

Plug in.

$$\lim_{x \to 9} \frac{\sqrt{x} - 3}{x - 9} = \lim_{x \to 9} \frac{1}{(\sqrt{x} + 3)} = \frac{1}{6}$$

## ▶ Practice

Evaluate the following limits.

**29.** $\displaystyle\lim_{x \to -2^+} \frac{(x - 6)(x + 2)}{(x + 2)(x + 1)}$

**30.** $\displaystyle\lim_{x \to 2} \frac{x - 2}{x^2 - 4}$

**31.** $\displaystyle\lim_{x \to 4} \frac{x^2 - 9}{x + 3}$

**32.** $\displaystyle\lim_{x \to 3} \frac{x^2 - 4x + 3}{x^2 + 2x - 15}$

**33.** $\displaystyle\lim_{x \to 3^-} \frac{x + 5}{x - 3}$

**34.** $\displaystyle\lim_{x \to 25} \frac{\sqrt{x} - 5}{(x - 25)(x + 1)}$

**35.** $\displaystyle\lim_{a \to 0} \frac{(x + a)^2 - x^2}{a}$

**36.** $\displaystyle\lim_{a \to 0} \frac{\sqrt{x + a} - \sqrt{x}}{a}$

# 6 ▶ Derivatives

S traight lines may be ideal to human beings, but most functions have curved graphs. This does not stop us from projecting straight lines on them! For example, at the point marked $x$ on the graph in Figure 6.1, the function is clearly increasing. However, exactly how fast is the function increasing at that point? Since "how fast" refers to a slope, we draw in the *tangent line,* the line straight through the point that heads in the same direction as the curve (see Figure 6.2). The slope of the tangent line tells us how fast the function is increasing at the given point.

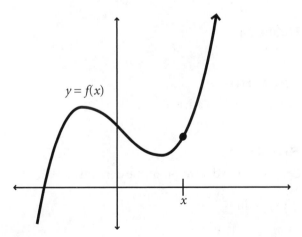

**Figure 6.1**

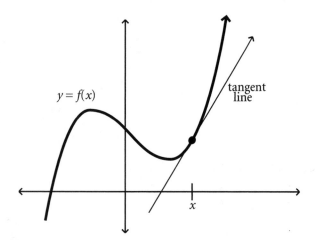

**Figure 6.2**

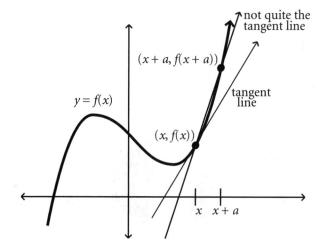

**Figure 6.3**

We can figure out the *y*-value of this point by plugging *x* into *f* and getting $(x, f(x))$. However, we can't get the slope of the tangent line when we have just one point. To get a second point, we go ahead a little further along the graph (see Figure 6.3). If we go ahead by distance *a*, the second point will have an *x*-value of $x + a$ and a *y*-value of $f(x + a)$.

Because this second point is on the curve and not on the tangent line, we get a line that is not quite the tangent line. Still, its slope will be close to the one we want, so we calculate as follows:

$$slope = \frac{f(x + a) - f(x)}{(x + a) - x} = \frac{f(x + a) - f(x)}{a}$$

To make things more accurate, we pick a second point that is closer to the first one by using a smaller *a*. This is depicted in Figure 6.4.

In fact, if we take the limit as *a* goes to zero, we will get the slope of the tangent line exactly. This is called the *derivative* of $f(x)$ and is written $f'(x)$.

$$f'(x) = \lim_{a \to 0} \frac{f(x + a) - f(x)}{a}$$

$$= \text{slope of the tangent line at point } (x, f(x))$$

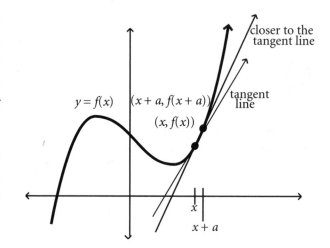

**Figure 6.4**

## Example

What is the derivative of $f(x) = x^2$?

## Solution

Start with the definition of the derivative.

$$f'(x) = \lim_{a \to 0} \frac{f(x + a) - f(x)}{a}$$

Use $f(x) = x^2$.

$$f'(x) = \lim_{a \to 0} \frac{(x + a)^2 - x^2}{a}$$

Multiply out and simplify.

$$f'(x) = \lim_{a \to 0} \frac{x^2 + 2ax + a^2 - x^2}{a}$$

Factor and simplify.

$$f'(x) = \lim_{a \to 0} \frac{(2x + a)a}{a}$$

Plug in for the limit.

$$f'(x) = \lim_{a \to 0} 2x + a = 2x$$

The derivative is $f'(x) = 2x$. This means that the slope at any point on the curve $y = x^2$ is exactly twice the $x$-coordinate. The situation at $x = -2$, $x = 0$, and $x = 1$ is shown in Figure 6.5.

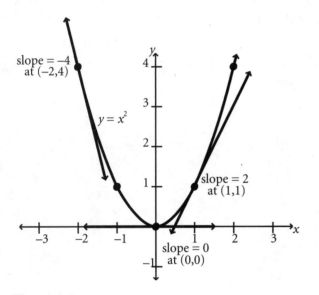

**Figure 6.5**

## Example

What is the slope of the line tangent to $g(x) = \sqrt{x}$ at $x = 9$?

## Solution

Start with the definition of the derivative.

$$g'(x) = \lim_{a \to 0} \frac{g(x + a) - g(x)}{a}$$

Use $g(x) = \sqrt{x}$.

$$g'(x) = \lim_{a \to 0} \frac{\sqrt{x + a} - \sqrt{x}}{a}$$

Rationalize the numerator.

$$g'(x) = \lim_{a \to 0} \left(\frac{\sqrt{x + a} - \sqrt{x}}{a}\right)\left(\frac{\sqrt{x + a} + \sqrt{x}}{\sqrt{x + a} + \sqrt{x}}\right)$$

Multiply and simplify.

$$g'(x) = \lim_{a \to 0} \frac{x + a + \sqrt{x} \cdot \sqrt{x + a} - \sqrt{x} \cdot \sqrt{x + a} - x}{a(\sqrt{x + a} + \sqrt{x})}$$

Simplify.

$$g'(x) = \lim_{a \to 0} \frac{a}{a(\sqrt{x + a} + \sqrt{x})}$$

Plug in to evaluate the limit.

$$g'(x) = \lim_{a \to 0} \frac{1}{\sqrt{x + a} + \sqrt{x}}$$
$$= \frac{1}{\sqrt{x + 0} + \sqrt{x}}$$
$$= \frac{1}{2\sqrt{x}}$$

The derivative of $g(x) = \sqrt{x}$ is thus $g'(x) = \frac{1}{2\sqrt{x}}$. This means that at $x = 9$, the slope of the tangent line is $g'(9) = \frac{1}{2\sqrt{9}} = \frac{1}{6}$. This is illustrated in Figure 6.6.

## Example

Find the equation of the tangent line to $h(x) = 2x^2 - 5x + 1$ at $x = 3$.

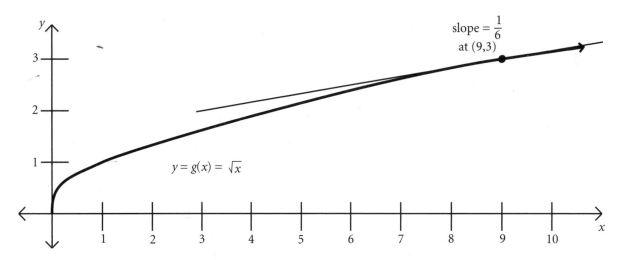

$$\text{slope} = \frac{1}{6}$$
$$\text{at } (9,3)$$

$y = g(x) = \sqrt{x}$

**Figure 6.6**

## Solution

To find the equation of the tangent line, we need a point and a slope. The $y$-value at $x = 3$ is $h(3) = 2(3)^2 - 5(3) + 1 = 4$, so the point is $(3,4)$. And to get the slope, we need the derivative. Start with the definition of the derivative.

$$h'(x) = \lim_{a \to 0} \frac{h(x + a) - h(x)}{a}$$

Use $h(x) = 2x^2 - 5x + 1$.

$$h'(x) = \lim_{a \to 0} \frac{2(x + a)^2 - 5(x + a) + 1 - (2x^2 - 5x + 1)}{a}$$

Multiply out and simplify.

$$h'(x) = \lim_{a \to 0} \frac{2x^2 + 4ax + 2a^2 - 5x - 5a + 1 - 2x^2 + 5x - 1}{a}$$

Factor out and simplify.

$$h'(x) = \lim_{a \to 0} \frac{(4x + 2a - 5)a}{a}$$

Evaluate the limit.

$$h'(x) = \lim_{a \to 0} 4x + 2a - 5 = 4x - 5$$

Thus, the derivative of $h(x) = 2x^2 - 5x + 1$ is $h'(x) = 4x - 5$. The slope at $x = 3$ is $h'(3) = 4(3) - 5 = 7$. The equation of the tangent line is therefore $y = 7(x - 3) + 4 = 7x - 17$. This is shown in Figure 6.7.

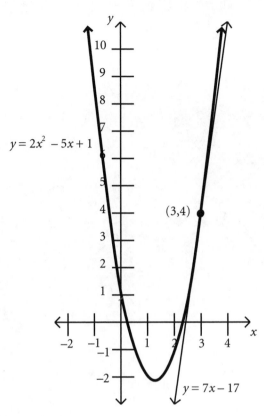

$y = 2x^2 - 5x + 1$

(3,4)

$y = 7x - 17$

**Figure 6.7**

**5.** Find the derivative of $f(x) = 3\sqrt{x}$.

**6.** If $k(x) = x^3$, then what is $k'(x)$?

**7.** Find the slope of $f(x) = 3x^2 + x$ at $x = 2$.

**8.** Where does the graph of $g(x) = x^2 - 4x + 1$ have a slope of 0?

**9.** Find the equation of the tangent line to $h(x) = 1 - x^2$ at $(2, -3)$.

**10.** What is the equation of the tangent line of $k(x) = 5x^2 + 2x$ at $x = 1$?

## ▶ Practice

**1.** Find the derivative of $f(x) = 8x + 2$.

**2.** If $h(x) = x^2 + 5$, then what is $h'(x)$?

**3.** Find the derivative of $g(x) = 10$.

**4.** What is the derivative of $g(x) = 3 - 5x$?

# Basic Rules of Differentiation

U sing the limit definition to find derivatives can be very tedious. Luckily, there are many shortcuts available. For example, if function $f$ is a constant, like $f(x) = 5$ or $f(x) = 18$, then $f'(x) = 0$. This can be proven for all constants $c$ at the same time in the following manner.

If:

$$f(x) = c$$

then:

$$f'(x) = \lim_{a \to 0} \frac{f(x + a) - f(x)}{a} = \lim_{a \to 0} \frac{c - c}{a} = \lim_{a \to 0} \frac{0}{a} = 0$$

All of the general rules in this chapter can be proven in such a manner, using the limit definition of the derivative, though we shall not bother to do so. The first rule is the Constant Rule, which says that if $f(x) = c$ where $c$ is a constant, then $f'(x) = 0$.

Before we go any further, a word needs to be said about notation. The concept of the derivative was discovered by both Isaac Newton and Gottfried Leibniz. Newton would put a dot over an object to represent its derivative, much like the way $f'(x)$ represents the derivative of $f(x)$. Leibniz would write the derivative of $y$ (where $x$ is the variable) as $\frac{dy}{dx}$. Newton's notation is certainly more convenient, but Leibniz's enables us

If $f(x) = c$ where $c$ is a constant, then $f'(x) = 0$.

And, using Leibnez's notation, if $c$ is a constant, then $\frac{d}{dx}(c) = 0$.

$$\frac{d}{dx}(x^n) = n \cdot x^{n-1}$$

to represent "take the derivative of something" as $\frac{d}{dx}$ (something). Thus, if $y = f(x)$, then $\frac{dy}{dx} = \frac{d}{dx}(f(x)) = f'(x)$. Using Leibniz's notation, the Constant Rule where $c$ is a constant is $\frac{d}{dx}(c) = 0$.

We will use both forms of the Constant Rule, depending on the situation. The next rule is the Power Rule, which is stated: $\frac{d}{dx}(x^n) = n \cdot x^{n-1}$. This rule says "multiply the exponent in front and then subtract one from it."

## Example

Differentiate $f(x) = x^2$.

## Solution

$f'(x) = 2x^{2-1} = 2x^1 = 2x$

## Example

Differentiate $y = x^8$.

## Solution

$\frac{dy}{dx} = 8x^7$

## Example

Differentiate $g(x) = \sqrt{x}$.

## Solution

To use the Power Rule, we need $g(x)$ expressed as $x$ raised to a power, or:

$g(x) = x^{\frac{1}{2}}$

$g'(x) = \frac{1}{2}x^{\frac{1}{2}-1} = \frac{1}{2}x^{-\frac{1}{2}} = \frac{1}{2} \cdot \frac{1}{\sqrt{x}} = \frac{1}{2\sqrt{x}}$

Notice how much easier it is to use the Power Rule to solve this problem than it was using the limit definition of the derivative in Lesson 6.

# The Constant Coefficient Rule

If a function has a constant multiplied in front, leave it while you take the derivative of the rest.

## Example

Differentiate $y = \dfrac{1}{x^2}$.

## Solution

Again, we have to rewrite $y$ as $x^{-2}$ so that it becomes $x$ raised to a power.

$$\frac{dy}{dx} = \frac{d}{dx}\left(\frac{1}{x^2}\right)$$

$$= \frac{d}{dx}(x^{-2}) = -2x^{-2-1} = -2x^{-3} = \frac{-2}{x^3}$$

## Example

Differentiate $y = \sqrt[3]{t}$.

## Solution

$$\frac{d}{dt}(\sqrt[3]{t}) = \frac{d}{dt}(t^{\frac{1}{3}}) = \frac{1}{3}t^{\frac{1}{3}-1} = \frac{1}{3}t^{-\frac{2}{3}} = \frac{1}{3t^{\frac{2}{3}}}$$

Notice that $\dfrac{d}{dt}$ means "take the derivative with respect to variable $t$." Usually, our variable is $x$, so the derivative of $y = f(x)$ is $\dfrac{dy}{dx} = f'(x)$, but sometimes, we have other variables. If $y = f(u)$, then $\dfrac{dy}{du} = f'(u)$ is the derivative with respect to $u$, for example.

## ▶ Practice

Differentiate each of the following.

**1.** $f(x) = x^5$

**2.** $y = x^7$

**3.** $g(u) = u^{-5}$

**4.** $h(x) = 8$

**5.** $y = t^4$

**6.** $y = x^{\frac{7}{5}}$

**7.** $f(x) = x^{100}$

**8.** $f(t) = -11$

**9.** $h(x) = x$

**10.** $y = x^{\frac{2}{3}}$

**11.** $g(x) = x^{-\frac{4}{5}}$

**12.** $k(x) = \sqrt[4]{x}$

**13.** $y = \sqrt{u}$

**14.** $y = \dfrac{1}{x}$

**15.** $f(x) = \dfrac{1}{\sqrt{x}}$

**16.** $g(x) = \dfrac{1}{x^3}$

If parts of a function are added together, differentiate the parts separately.

## ▶ The Constant Coefficient Rule

The Constant Coefficient Rule is stated as follows: If a function has a constant multiplied in front, leave it while you take the derivative of the rest. This means that because $\frac{d}{dx}(x^8) = 8x^7$, then the derivative of $5x^8$ is $5 \cdot (8x^7) = 40x^7$. Just imagine that the constant steps aside and waits while you differentiate the rest.

### Examples

Differentiate the following.

$$f(x) = 11x^4$$

$$y = 10x^2$$

$$g(x) = 3\sqrt{x} = 3x^{\frac{1}{2}}$$

$$h(t) = \frac{4}{t^6} = 4t^{-6}$$

$$y = 12x$$

$$k(u) = \frac{15\sqrt[3]{u}}{4} = \frac{15}{4}u^{\frac{1}{3}}$$

$$A(r) = \pi r^2$$

### Solutions

$$f'(x) = 44x^3$$

$$\frac{dy}{dx} = 20x$$

$$g'(x) = \frac{3}{2}x^{-\frac{1}{2}} = \frac{3}{2\sqrt{x}}$$

$$h'(t) = -24t^{-7} = -\frac{24}{t^7}$$

$$\frac{dy}{dx} = 12$$

$$k'(u) = \frac{5}{4}u^{-\frac{2}{3}} = \frac{5}{4u^{\frac{2}{3}}}$$

$$A'(r) = 2\pi r$$

In that last example problem, don't forget that $\pi$ is a constant, and thus $2\pi r$ should be treated just as $20r$ or $712r$ would.

Remember that $x^0 = 1$. This means that a constant function such as $f(x) = 5$ could also be written $f(x) = 5 \cdot 1 = 5x^0$. Using both the Power Rule and the Constant Coefficient Rule, it would look like this:

$$f'(x) = 5(0 \cdot x^{0-1}) = 5 \cdot 0 \cdot x^{-1} = 0$$

This shows that the Constant Rule really isn't necessary, because the Power Rule and the Constant Coefficient Rule together say that the derivative of a constant is zero.

## ▶ The Additive Rule

Next, we will examine the Additive Rule, which says that if parts of a function are added together, differentiate the parts separately. We know that

$\frac{d}{dx}(10x^2) = 20x$ and $\frac{d}{dx}(12x) = 12$. The Additive Rule then says that if $y = 10x^2 + 12x$, then $\frac{dy}{dx} = \frac{d}{dx}(10x^2 + 12x) = \frac{d}{dx}(10x^2) + \frac{d}{dx}(12x) = 20x + 12$. Because the $10x^2$ and $12x$ are added together, we differentiate them separately.

## Example
Differentiate $f(x) = 4x^5 + 30x^2$.

## Solution
$f'(x) = 20x^4 + 60x$

## Example
Differentiate $g(x) = x^3 - 4x^2$.

## Solution
This can be rewritten as addition:

$g(x) = x^3 + (-4)x^2$

thus:

$g'(x) = 3x^2 + (-4) \cdot 2x = 3x^2 - 8x$.

The previous example shows that the Additive Rule applies to cases of subtraction as well.

## Examples
Differentiate the following.

$y = \sqrt{x} + 4 = x^{\frac{1}{2}} + 4$

$h(x) = 8x^5 + 10x^4 - 3x^3 + 7x^2 - 5x + 4$

$k(t) = 3t^{\frac{4}{5}} + \frac{2}{t} + 11 = 3t^{\frac{4}{5}} + 2t^{-1} + 11$

## Solutions

$\frac{dy}{dx} = \frac{1}{2\sqrt{x}} + 0 = \frac{1}{2\sqrt{x}}$

$h'(x) = 40x^4 + 40x^3 - 9x^2 + 14x - 5$

$k'(t) = \frac{12}{5}t^{-\frac{1}{5}} - 2t^{-2} = \frac{12}{5\sqrt[5]{t}} - \frac{2}{t^2}$

## ▶ Practice

Differentiate the following.

**17.** $y = 3x^7$

**18.** $f(x) = \frac{-3}{x^{10}}$

**19.** $V(r) = \frac{4}{3}\pi r^3$

**20.** $g(t) = \frac{12t^4}{5}$

**21.** $k(x) = 1 - x^2$

**22.** $y = 4t^3 - 8t + 70$

**23.** $f(x) = 8x^3 + 3x^2$

**24.** $y = x^2 - 3x + 5$

**25.** $s(t) = -16t^2 + 5t + 200$

**26.** $F(x) = 6x^{100} + 10x^{50} - 4x^{25} + 2x^{10} - 9$

**27.** $g(x) = 3x^{\frac{1}{6}} + 5x^3$

**28.** $h(u) = u^5 + 4u^4 - 7u^3 - 2u^2 + 8u - 2$

**29.** $y = 3 + \frac{2}{x} + \frac{1}{x^2}$

**30.** $y = u^2 - u^{-2}$

**31.** $f(x) = 4x^2 - 8x + 5 + \dfrac{3}{x}$

**32.** $y = 4\sqrt{x} + 9\sqrt[3]{x}$

The derivative of the derivative is called the *second derivative*. The derivative of that is the *third derivative,* and so on. This is where Newton's notation really shines. If $y = f(x)$, then the derivative is $\dfrac{dy}{dx} = f'(x)$ and the second derivative is $\dfrac{d^2y}{dx^2} = f''(x)$. The third derivative is $\dfrac{d^3y}{dx^3} = f'''(x)$, and the tenth derivative, for example, is $\dfrac{d^{10}y}{dx^{10}} = f^{(10)}(x)$. We put the 10 in parentheses like that because counting the ten primes in $f''''''''''(x)$ gets ridiculous.

### Example
Find the first three derivatives of $y = \sqrt{x}$.

### Solution

$y = x^{\frac{1}{2}}$

$\dfrac{dy}{dx} = \dfrac{1}{2}x^{-\frac{1}{2}}$

$\dfrac{d^2y}{dx^2} = -\dfrac{1}{4}x^{-\frac{3}{2}}$

$\dfrac{d^3y}{dx^3} = \dfrac{3}{8}x^{-\frac{5}{2}}$

When working on multiple derivatives like this, it makes sense to leave the exponents negative and fractional.

### Example
Find all the derivatives of $f(x) = x^3 - 4x^2 + 5x - 7$.

### Solution

$f(x) = x^3 - 4x^2 + 5x - 7$

$f'(x) = 3x^2 - 8x + 5$

$f''(x) = 6x - 8$

$f'''(x) = 6$

$f''''(x) = 0$

All of the subsequent derivatives will also be zero, so we can write

$f^{(n)}(x) = 0$ for $n \geq 4$.

### ▶ Practice

**33.** Find the first four derivatives of $f(x) = \dfrac{1}{x}$.

**34.** Find the second derivative of
$s(t) = -16t^2 + 20t + 150$.

**35.** Find the third derivative of
$y = 10x^4 - 7x^3 + 6x - 1$.

**36.** Find the first three derivatives of $y = 6\sqrt[3]{t}$.

# 8 ▶ Rates of Change

I t is useful to contemplate slopes in practical situations. For example, suppose the following graph in Figure 8.1 is for $y = f(x)$, a function that gives the price $y$ for various amounts $x$ of cheese. Because the straight line goes through the points (1 lb.,$2) and (2 lbs.,$4), the slope $= \dfrac{4 - 2}{2 \text{ lbs.} - 1 \text{ lb.}} = \dfrac{2}{1 \text{ lb.}} = \$2$ per pound.

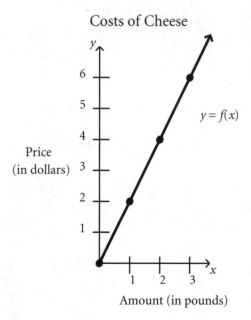

Costs of Cheese

$y = f(x)$

Price (in dollars)

Amount (in pounds)

**Figure 8.1**

The slope is therefore the *rate* at which the cheese is sold, in dollars per pound. Because slope = $\frac{y\text{-}change}{x\text{-}change}$, a slope will always be a rate measured in *y*-units per *x*-unit.

For example, suppose a passenger on a bus writes down the exact time she passes each highway mile marker. She then sketches the graph shown in Figure 8.2 of the bus's position on the highway over time. The slope at any point on this graph will be measured in *y*-units per *t*-unit, or miles per hour. The steepness of the slope represents the speed of the bus.

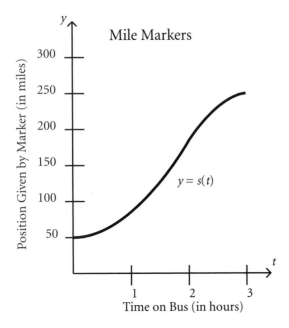

**Figure 8.2**

► **Practice**

For each of the following four graphs, give the rate that a slope represents.

**1.**

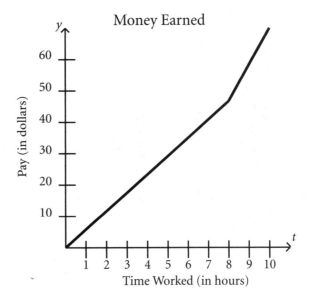

**2.**

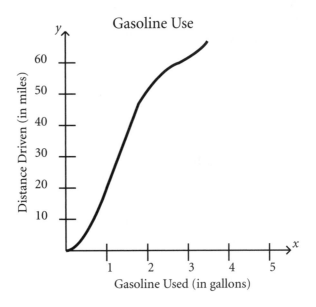

**3.**

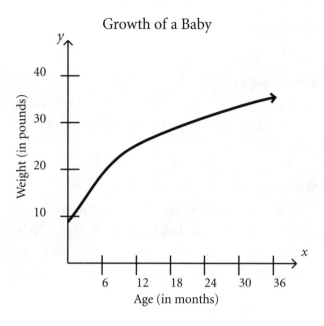

Growth of a Baby

**4.**

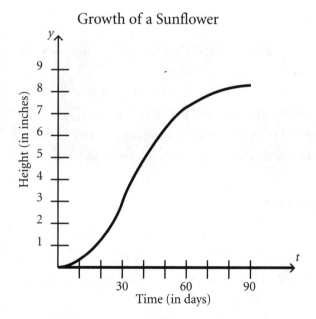

Growth of a Sunflower

Because the derivative of a function gives the slope of its tangent lines, these practice problems show that the derivative of a function gives its *rate of change*.

An excellent example comes from position functions. A *position function* $s(t)$ states where an object is at any given time. The derivative $s'(t)$ states the rate at which that object's position is changing—that is, the speed or *velocity* of the function. Thus, $s'(t) = v(t)$. The second derivative $s''(t) = v'(t)$ tells how the velocity is changing, or the *acceleration*. Thus, $s''(t) = v'(t) = a(t)$ where $s(t)$ is the position function, $v(t)$ is the velocity function, and $a(t)$ is the acceleration function.

## Example

Suppose an object rolls along beside a tape measure so that after $t$ seconds, it is next to the inch marked $s(t) = 4t^2 + 8t + 5$. Where is the object after 1 second? After 3 seconds? What is the velocity function? How fast is the object moving after 2 seconds? What is the acceleration function?

## Solution

The position function $s(t) = 4t^2 + 8t + 5$ tells us where the object is. After 1 second, the object is next to the $s(1) = 17$-inch mark on the tape measure. After 3 seconds, the object is at the $s(3) = 65$-inch mark.

The velocity function is $v(t) = s'(t) = 8t + 8$. Thus, after 2 seconds, the object is moving at the rate of $v(2) = 24$ inches per second. Do realize that this velocity of 24 inches per second is an *instantaneous velocity*, the speed just at a single moment. If a car's speedometer reads 60 miles per hour, this does not mean that it will drive for 60 miles or even for a full hour. The car might speed up, slow down, or stop. However, at that instant, the car is traveling at a rate that, if unchanged, will take it 60 miles in one hour. A derivative is always an instantaneous rate, telling you the slope at a particular point, but not making any promises about what will happen next.

The acceleration function is $a(t) = v'(t) = s''(t) = 8$. Because this is a constant, it tells us that the object increases in speed by 8 inches per second every second.

The most popular example of constant acceleration is gravity, which accelerates objects downward by $32\frac{ft}{sec}$ every second. Because of this, an object thrown with a velocity of $b$ feet per second from a height of $h$ feet above the ground will have (after $t$ seconds) a height of $s(t) = -16t^2 + bt + h$ feet.

The starting time is $t = 0$, at which point the object is $s(0) = h$ feet off the ground, the correct initial height. The velocity function is $v(t) = s'(t) = -32t + b$. At the starting time $t = 0$, the velocity is $v(0) = b$, the desired initial velocity. The function $v(t) = -32t + b$ means that 32 feet per second are subtracted from the initial velocity $b$ every second. The acceleration function is $a(t) = v'(t) = s''(t) = -32$. This is the desired constant acceleration.

### Example

Suppose a brick is thrown upward at $10\frac{ft}{sec}$ from a 150-foot rooftop. What are its position, velocity, and acceleration functions?

### Solution

Because the initial velocity is $b = 10\frac{ft}{sec}$ and the initial height is $h = 150$ feet, the position function is $s(t) = -16t^2 + 10t + 150$. The velocity function is $v(t) = s'(t) = -32t + 10$. The acceleration is $a(t) = -32$, a constant 32 feet per second downward

each second. The negative sign indicates that gravity is acting to decrease the height of the brick, pulling it downward.

### Example

Suppose a rock is dropped from a 144-foot tall bridge. When will the rock hit the water? How fast will it be going then?

### Solution

Because the rock is dropped, the initial velocity is $b = 0$. The initial height is $h = 144$. Thus, $s(t) = -16t^2 + 144$ gives the height function. The rock will hit the water (have a height of zero) when:

$$-16t^2 + 144 = 0$$

$$144 = 16t^2$$

$$t = \pm 3$$

And because $-3$ seconds doesn't make any sense, the rock will hit after 3 seconds.

The velocity function is $v(t) = s'(t) = -32t$; therefore, the rock will have a velocity of $v(3) = -96$ after 3 seconds. It will be traveling at a rate of 96 feet per second downward when it hits the water.

### Example

If $p(t) = \frac{t^2}{10} - 80t + 50{,}000$ gives the value, in thousands of dollars, of a start-up company after $t$ days, then how fast is its value changing after 30 days? After 500 days?

## Solution

The derivative $p'(t) = \frac{t}{5} - 80$ gives the rate of change in value, measured in thousands of dollars per day. After 30 days, $p'(30) = -74$, so the company will be losing $74,000 of value every day. After 500 days, $p'(500) = 20$, so the company will be gaining value at the instantaneous rate of $20,000 a day.

## ▶ Practice

5. The height of a tree after $t$ years is $h(t) = 30 - \frac{25}{t}$ feet when $t \geq 1$. How fast is the tree growing after 5 years?

6. The level of a river $t$ days after a heavy rainstorm is $L(t) = -t^2 + 8t + 26$ feet. How fast is the river's level changing after 7 days?

7. When a company makes and sells $x$ cars, its profit is $P(x) = \frac{x^3}{10} - 60x^2 + 9,000x$ dollars. How fast is its profit changing when the company makes 50 cars? Should the company make more cars?

8. When a container is made $x$ inches wide, it costs $C(x) = 0.8x^2 + \frac{24}{x}$ dollars to make. How is the cost changing when $x = 3$ inches? Would it be cheaper to increase or decrease the width?

9. An electron in a particle accelerator is $s(t) = t^3 + 2t^2 + 10t$ meters from the start after $t$ seconds. Where is it after 3 seconds? How fast is it moving then? How fast is it accelerating then?

10. A brick is dropped from 64 feet off the ground. What is its position function? What is its velocity function? What is its acceleration? When will it hit the ground? How fast will it be traveling then?

11. A bullet is fired upward at 800 feet per second from the ground. How high is it when it stops rising and starts to fall?

12. A rock is thrown 10 feet per second down a 1,000-foot cliff. How far has it gone down in the first 4 seconds? How fast is it traveling then?

## ▶ Derivatives of Sine and Cosine

It is by examining rates and slopes that we can find the derivative of $\sin(x)$. Look at the slopes at various points on its graph in Figure 8.2. It appears that the derivative function of $\sin(x)$ must oscillate between $-1$ and 1, and must go through the following points (see Figure 8.3). The function $\cos(x)$ is exactly such an oscillating function (see Figure 8.4). This leads us to conclude $\frac{d}{dx}(\sin(x)) = \cos(x)$.

A similar study of the slopes of $\cos(x)$ would show that $\frac{d}{dx}(\cos(x)) = -\sin(x)$. The slopes of the cosine function are not the values of the sine function, but rather their exact negatives.

### Examples

Differentiate the following examples.

$f(x) = 5\sin(x) + 4x^2$

$y = 2 + \cos(t)$

$g(x) = \sin(x) - \cos(x)$

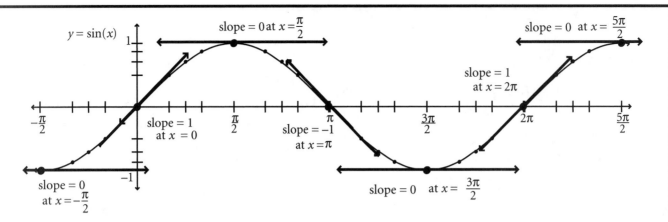

**Figure 8.2**

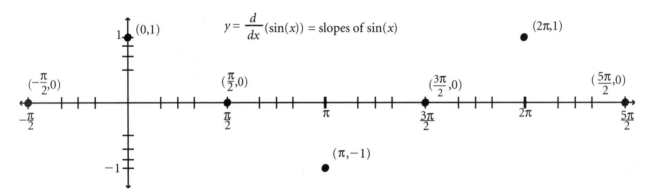

**Figure 8.3**

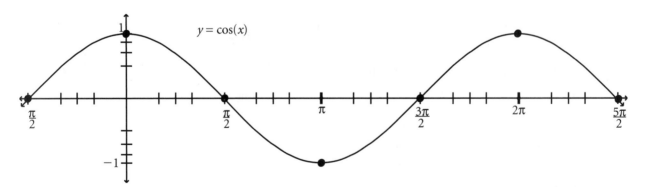

**Figure 8.4**

## Solutions

$$f'(x) = 5\cos(x) + 8x$$

$$\frac{dy}{dt} = -\sin(t)$$

$$g'(x) = \cos(x) + \sin(x)$$

## ▶ Practice

Differentiate the following practice problems.

**13.** $y = 4x^5 + 10\cos(x) + 3$

**14.** $f(t) = 3\sin(t) + \dfrac{2}{t}$

**15.** $g(x) = 8x + 3 - \cos(x)$

**16.** $r(\theta) = \dfrac{1}{2}\sin(\theta) + \dfrac{1}{2}\cos(\theta)$

**17.** $h(x) = \cos(x) + \cos(5)$

**18.** Find the equation of the tangent line to
$f(x) = \sin(x) + \cos(x)$ at $x = \dfrac{\pi}{2}$.

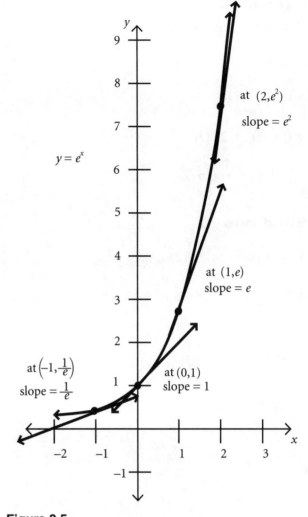

**Figure 8.5**

## ▶ Derivatives of the Exponential and Natural Logarithm Functions

The reason why the nicest exponential function is $e^x$ where $e = 2.71828\ldots$ is because this makes for the following very nice derivative:

$$\frac{d}{dx}(e^x) = e^x$$

It is only with this exact base that the derivative of the exponential function is itself (see Figure 8.5).

As for the inverse function $\ln(x)$, the natural logarithm is written as follows:

$$\frac{d}{dx}(\ln(x)) = \frac{1}{x}$$

Lesson 11 will contain a proof for this.

## Examples

Differentiate the following.

$f(x) = 4e^x$

$y = 10e^x + 10$

$g(t) = 3e^t + 2\ln(t)$

$y = 8\ln(u) - e^u + 7u$

## Solutions

$f'(x) = 4e^x$

$\dfrac{dy}{dx} = 10e^x$

$g'(t) = 3e^t + \dfrac{2}{t}$

$\dfrac{dy}{du} = \dfrac{8}{u} - e^u + 7$

## ▶ Practice

Differentiate the following.

**19.** $f(x) = 1 + x + x^2 + x^3 + e^x$

**20.** $g(t) = 12\ln(t) + t^2 + 4$

**21.** $y = \cos(x) - 10e^x + 8x$

**22.** $h(x) = \sqrt{x} - 8\ln(x)$

**23.** $k(u) = 3x^{\frac{5}{2}} + 5e^x + 11$

**24.** Find the second derivative of $f(x) = e^x + \ln(x)$.

**25.** Find the 100th derivative of $g(x) = 3e^x$.

**26.** What is the slope of the tangent line to $f(x) = \ln(x)$ at $x = 10$?

LESSON

9 ▶

# The Product and Quotient Rules

## ▶ The Product Rule

When a function consists of parts that are added together, it is easy to take its derivative: Simply take the derivative of each part and add them together. We are inclined to try the same trick when the parts are multiplied together, but it does not work.

For example, we know that $\frac{d}{dx}(x^2) = 2x$ and $\frac{d}{dx}(x^3) = 3x^2$. The derivative of their product is

$\frac{d}{dx}(x^2 \cdot x^3) = \frac{d}{dx}(x^5) = 5x^4$. This shows that the derivative of a product is *not* the product of the derivatives:

$$5x^4 = \frac{d}{dx}(x^2 \cdot x^3) \neq \frac{d}{dx}(x^2) \cdot \frac{d}{dx}(x^3) = (2x) \cdot (3x^2) = 6x^3$$

Instead, we take the derivative of each part, multiply by the other part *left alone,* and add the results together:

$$\frac{d}{dx}(x^2 \cdot x^3) = \frac{d}{dx}(x^2) \cdot x^3 + \frac{d}{dx}(x^3) \cdot x^2 = (2x) \cdot x^3 + (3x^2) \cdot x^2 = 5x^4$$

This time, we *did* get the correct answer.

# The Product Rule

The Product Rule can be stated "the derivative of the first times the second, plus the derivative of the second times the first." It can be proven directly from the limit definition of the derivative, but only with a few tricks and a lot of algebra. The Product Rule is given as follows:

$$\frac{d}{dx}(f(x) \cdot g(x)) = f'(x) \cdot g(x) + g'(x) \cdot f(x)$$

## Example
Differentiate $y = x^3 \sin(x)$.

## Solution
Here, the "first part" is $x^3$ and the "second part" is $\sin(x)$. Thus, by using the Product Rule, $\frac{d}{dx}(x^3\sin(x)) = \frac{d}{dx}(x^3) \cdot \sin(x) + \frac{d}{dx}(\sin(x)) \cdot x^3 = 3x^2\sin(x) + \cos(x) \cdot x^3$. This could be simplified as $\frac{dy}{dx} = x^2(3\sin(x) + x\cos(x))$, but that's not really all that's necessary.

## Example
Differentiate $f(x) = \ln(x) \cdot \cos(x)$.

## Solution

$$f'(x) = \frac{d}{dx}(\ln(x)) \cdot \cos(x) + \frac{d}{dx}(\cos(x)) \cdot \ln(x)$$

$$= \frac{1}{x} \cdot \cos(x) - \sin(x) \cdot \ln(x)$$

Thus, the derivative is:

$$f'(x) = \frac{\cos(x)}{x} - \ln(x) \cdot \sin(x)$$

## Example

Differentiate $g(x) = 5x^7 \cdot e^x$.

## Solution

$$g'(x) = \frac{d}{dx}(5x^7) \cdot e^x + \frac{d}{dx}(e^x) \cdot 5x^7$$

$$= 35x^6 \cdot e^x + e^x \cdot 5x^7 = 5x^6e^x(7 + x)$$

Using the product rule with $e^x$ can be a little bit confusing because there is no difference between the derivative of $e^x$ and $e^x$ "left alone." Still, if you write everything out, the correct answer should fall into place, even if it looks weird.

## Example
Differentiate $y = t^2\ln(t)$.

## Solution

$$\frac{dy}{dt} = \frac{d}{dt}(t^2)\ln(t) + \frac{d}{dt}(\ln(t))t^2$$

$$= 2t \cdot \ln(t) + \frac{1}{t} \cdot t^2$$

$$= 2t \cdot \ln(t) + t$$

$$= t(2 + \ln(t))$$

## Example
Differentiate $y = x^5\sin(x)\cos(x)$.

## Solution
We'll use the Product Rule with $x^5$ as the first part and $\sin(x)\cos(x)$ as the second part. However, in taking the derivative of $\sin(x)\cos(x)$, we'll have to use the

$$\frac{d}{dx}\left(\frac{f(x)}{g(x)}\right) = \frac{f'(x)g(x) - g'(x)f(x)}{(g(x))^2}$$

Product Rule a second time. It might get messy, but it will work if everything is written down carefully.

$$\frac{dy}{dx} = \frac{d}{dx}(x^5) \cdot \sin(x)\cos(x) + \frac{d}{dx}(\sin(x)\cos(x)) \cdot x^5$$

$$\frac{dy}{dx} = 4x^5\sin(x)\cos(x) + \left[\frac{d}{dx}(\sin(x)) \cdot \cos(x) + \frac{d}{dx}(\cos(x)) \cdot \sin(x)\right] \cdot x^5$$

$$\frac{dy}{dx} = 4x^5\sin(x)\cos(x) + \left[\cos(x) \cdot \cos(x) - \sin(x) \cdot \sin(x)\right] \cdot x^5$$

$$\frac{dy}{dx} = 4x^5\sin(x)\cos(x) + (\cos^2(x) - \sin^2(x)) \cdot x^5$$

## ▶ Practice

Differentiate the following.

**1.** $f(x) = x^2\cos(x)$

**2.** $y = 8t^3e^t$

**3.** $y = \sin(x)\cos(x)$

**4.** $g(x) = 3x^2\ln(x) - 5x^4 + 10$

**5.** $h(u) = ue^u - e^u$

**6.** $k(x) = \sin(x) + x^4 - x^2\sin(x)$

**7.** $y = 8\ln(x)\sin(x) + \cos(x)$

**8.** $h(t) = t\sin(t) - t\cos(t)$

**9.** $y = 5x^3 - x\ln(x)$

**10.** $f(x) = \sin^2(x) = \sin(x) \cdot \sin(x)$

**11.** $y = xe^x\sin(x)$

**12.** $g(x) = 3x^4\ln(x)\cos(x)$

**13.** What is the slope of the tangent line to $f(x) = x^2e^x + x + 2$ at $(0,2)$?

**14.** Find the equation of the tangent line to $y = x\sin(x)$ at $x = \pi$.

## ▶ The Quotient Rule

The Quotient Rule for functions where the parts are divided is even more complicated than the Product Rule. The Quotient Rule can be stated:

$$\frac{d}{dx}\left(\frac{f(x)}{g(x)}\right) = \frac{f'(x)g(x) - g'(x)f(x)}{(g(x))^2}$$

Just as with the Product Rule, each part is differentiated and multiplied by the other part. Here, however, they are subtracted, so it matters which one is differentiated first. It is important to start with the derivative of the top.

## Example

Differentiate $y = \dfrac{x^5 - 3x^2 + 1}{\cos(x)}$.

## Solution

Here, the top part is $x^5 - 3x^2 + 1$ and the bottom part is $\cos(x)$. Therefore, by the Quotient Rule:

$$\frac{dy}{dx} = \frac{\frac{d}{dx}(x^5 - 3x^2 + 1) \cdot \cos(x) -}{(\cos(x))^2}$$

$$\frac{\frac{d}{dx}(\cos(x)) \cdot (x^5 - 3x^2 + 1)}{(\cos(x))^2}$$

$$\frac{dy}{dx} =$$

$$\frac{(5x^4 - 6x) \cdot \cos(x) - (-\sin(x)) \cdot (x^5 - 3x^2 + 1)}{\cos^2(x)}$$

$$\frac{dy}{dx} =$$

$$\frac{(5x^4 - 6x) \cdot \cos(x) + \sin(x) \cdot (x^5 - 3x^2 + 1)}{\cos^2(x)}$$

## Example

Differentiate $f(x) = \dfrac{x^3}{10x^2 - 1}$.

## Solution

$$f'(x) = \frac{\frac{d}{dx}(x^3) \cdot (10x^2 - 1) - \frac{d}{dx}(10x^2 - 1) \cdot x^3}{(10x^2 - 1)^2}$$

$$f'(x) = \frac{(3x^2) \cdot (10x^2 - 1) - (20x) \cdot x^3}{(10x^2 - 1)^2}$$

$$f'(x) = \frac{30x^4 - 3x^2 - 20x^4}{(10x^2 - 1)^2} = \frac{10x^4 - 3x^2}{(10x^2 - 1)^2}$$

## Example

Differentiate $y = \dfrac{x^2\sin(x)}{\ln(x)}$.

## Solution

Here, the Product Rule is necessary to differentiate the top.

$$\frac{dy}{dx} = \frac{\frac{d}{dx}(x^2\sin(x)) \cdot \ln(x) - \frac{d}{dx}(\ln(x)) \cdot x^2\sin(x)}{(\ln(x))^2}$$

$$\frac{dy}{dx} = \frac{\left[\frac{d}{dx}(x^2) \cdot \sin(x) + \frac{d}{dx}(\sin(x)) \cdot x^2\right] \cdot \ln(x) - \frac{1}{x} \cdot x^2\sin(x)}{(\ln(x))^2}$$

$$\frac{dy}{dx} = \frac{\left[2x \cdot \sin(x) + \cos(x) \cdot x^2\right] \cdot \ln(x) - x\sin(x)}{(\ln(x))^2}$$

## Example

Differentiate $y = \dfrac{\ln(t)}{t}$.

## Solution

$$\frac{dy}{dt} = \frac{\frac{d}{dt}(\ln(t)) \cdot t - \frac{d}{dt}(t) \cdot \ln(t)}{t^2}$$

$$\frac{dy}{dt} = \frac{\frac{1}{t} \cdot t - 1 \cdot \ln(t)}{t^2}$$

$$\frac{dy}{dt} = \frac{1 - \ln(t)}{t^2}$$

Some people remember the Quotient Rule as

$$\frac{d}{dx}\left(\frac{HI}{HO}\right) = \frac{HO \cdot d(HI) - HI \cdot d(HO)}{HO \cdot HO}$$

just so they can say "HI d'HO" and "HO HO," but they're silly.

## ▶ Practice

Differentiate the following.

**15.** $h(x) = \dfrac{x^3 + 10x - 7}{3x^2 + 5x + 2}$

**16.** $y = \dfrac{4e^t + t}{t^3 + 2t + 1}$

**17.** $f(x) = \dfrac{x + \ln(x)}{e^x - 1}$

**18.** $y = \dfrac{x^5}{\ln(x)}$

**19.** $f(x) = \dfrac{x^2 - 1}{x^2 + 1}$

**20.** $g(t) = \dfrac{t^3}{5\sin(t)}$

**21.** $y = \dfrac{x + 1}{x - 1}$

**22.** $g(u) = \dfrac{\sin(u)}{u^3 - e^u}$

**23.** $y = \dfrac{x^2 + 2x + e^x}{\sin(x) + 1}$

**24.** $h(t) = \dfrac{\ln(t) + t}{t^2}$

**25.** $y = \dfrac{x\ln(x)}{e^x}$

**26.** $f(x) = \dfrac{x^2 e^x}{\cos(x)}$

**27.** Find the second derivative of $y = \dfrac{x + 1}{x - 4}$.

**28.** What is the slope of the tangent line to
$$f(x) = \frac{x^2 + x}{x + 5} \text{ at } x = 5?$$

## ▶ Derivatives of Trigonometric Functions

With the Quotient Rule, we can find the derivatives of all of the rest of the trigonometric functions.

**Example**
Differentiate $y = \tan(x)$.

**Solution**
Use $\tan(x) = \dfrac{\sin(x)}{\cos(x)}$.

$$\frac{dy}{dx} = \frac{d}{dx}(\tan(x)) = \frac{d}{dx}\left(\frac{\sin(x)}{\cos(x)}\right)$$

Differentiate with the Quotient Rule.
$$\frac{dy}{dx} = \frac{\cos(x) \cdot \cos(x) - (-\sin(x)) \cdot \sin(x)}{\cos^2(x)}$$

Simplify.
$$\frac{dy}{dx} = \frac{\cos^2(x) + \sin^2(x)}{\cos^2(x)}$$

Use $\sin^2(x) + \cos^2(x) = 1$.

$$\frac{dy}{dx} = \frac{1}{\cos^2(x)}$$

Use $\sec(x) = \dfrac{1}{\cos(x)}$.

$$\frac{dy}{dx} = \sec^2(x)$$

Thus:

$$\frac{d}{dx}(\tan(x)) = \sec^2(x)$$

**Example**

Differentiate $y = \sec(x)$.

**Solution**

Use $\sec(x) = \dfrac{1}{\cos(x)}$.

$$\frac{dy}{dx} = \frac{d}{dx}(\sec(x)) = \frac{d}{dx}\left(\frac{1}{\cos(x)}\right)$$

Differentiate with the Quotient Rule.

$$\frac{dy}{dx} = \frac{0 \cdot \cos(x) - (-\sin(x)) \cdot 1}{\cos^2(x)}$$

Simplify.

$$\frac{dy}{dx} = \frac{\sin(x)}{\cos^2(x)} = \frac{1}{\cos(x)} \cdot \frac{\sin(x)}{\cos(x)}$$

Use $\sec(x) = \dfrac{1}{\cos(x)}$ and $\tan(x) = \dfrac{\sin(x)}{\cos(x)}$.

$$\frac{dy}{dx} = \sec(x)\tan(x)$$

Thus:

$$\frac{d}{dx}(\sec(x)) = \sec(x)\tan(x)$$

## ▶ Practice

Differentiate the following.

**29.** $y = \csc(x)$

**30.** $y = \cot(x)$

**31.** $f(x) = x\tan(x)$

**32.** $g(x) = \dfrac{\sqrt{x}}{\sec(x)}$

# 10 ▶ Chain Rule

We have found how to take derivatives of functions that are added, subtracted, multiplied, and divided. Next, we will cover how to work with a function that is put inside another simply by composition.

For example, it would be difficult to multiply out $f(x) = (x^3 + 10x + 4)^5$ just to take the derivative. Instead, notice that $f(x)$ looks like $g(x) = x^3 + 10x + 4$ put inside $h(x) = x^5$. Therefore, in terms of composition, $f(x) = h \circ g(x) = h(g(x))$.

The trick to differentiating composed functions is to take the derivative of the outermost layer first, while leaving the inner part alone, then multiplying by the derivative of the inside.

The Chain Rule can be stated as follows:

$$\frac{d}{dx}(h(g(x))) = h'(g(x)) \cdot g'(x)$$

If this is confusing, try stating the Chain Rule in this way:

$$\frac{d}{dx}h(\text{something}) = h'(\text{something}) \cdot \frac{d}{dx}(\text{something})$$

$$\frac{d}{dx}(h(g(x))) = h'(g(x)) \cdot g'(x) \quad \text{or} \quad \frac{d}{dx}(h(\text{something})) = h'(\text{something}) \cdot \frac{d}{dx}(\text{something})$$

The usual key to figuring out what is inside and what is outside is to watch the parentheses. Imagine that the parentheses form the layers of an onion, and that you must peel (differentiate) the outermost layers before reaching the inside.

## Example

Differentiate $f(x) = (x^3 + 10x + 4)^5$.

## Solution

Here, $f(x) = (\text{something})^5$ where the something $= x^3 + 10x + 4$. Because $\frac{d}{dx}(x^5) = 5x^4$, the Chain Rule states the following:

$$f'(x) = 5(\text{something})^4 \cdot \frac{d}{dx}(\text{something})$$

$$f'(x) = 5(x^3 + 10x + 4)^4 \cdot \frac{d}{dx}(x^3 + 10x + 4)$$

$$f'(x) = 5(x^3 + 10x + 4)^4 \cdot (3x^2 + 10)$$

## Example

Differentiate $g(x) = \sin(8x^4 + 3x^2 - 2x + 1)$.

## Solution

Here, the function is essentially sin(something) where the "something" $= 8x^4 + 3x^2 - 2x + 1$. The derivative of sine is cosine, so:

$$g'(x) = \cos(\text{something}) \cdot \frac{d}{dx}(\text{something})$$

$$g'(x) = \cos(8x^4 + 3x^2 - 2x + 1) \cdot$$
$$\frac{d}{dx}(8x^4 + 3x^2 - 2x + 1)$$

$$g'(x) = \cos(8x^4 + 3x^2 - 2x + 1) \cdot$$
$$(32x^3 + 6x - 2)$$

## Example

Differentiate $y = \cos^3(x)$.

## Solution

This is tricky because of the laziness of mathematicians who like to skimp on parentheses. It might look like the "outside" function is cos(something), but it is actually $y = \cos^3(x) = (\cos(x))^3$. Thus, this function is really $(\text{something})^3$.

$$\frac{dy}{dx} = 3(\text{something})^2 \cdot \frac{d}{dx}(\text{something})$$

$$\frac{dy}{dx} = 3(\cos(x))^2 \cdot \frac{d}{dx}(\cos(x))$$

$$\frac{dy}{dx} = 3(\cos(x))^2 \cdot (-\sin(x))$$

$$\frac{dy}{dx} = -3\cos^2(x)\sin(x)$$

## Example

Differentiate $y = \cos(x^3)$.

It is important that the "something" in the parentheses appear somewhere in the derivative, just as it does in the original function. If it doesn't appear, then a mistake has been made.

## Solution

In this example, our function is $\cos(\text{something})$. Because $\frac{d}{dx}(\cos(x)) = -\sin(x)$, the Chain Rule states that

$$\frac{dy}{dx} = -\sin(\text{something}) \cdot \frac{d}{dx}(\text{something})$$

$$\frac{dy}{dx} = -\sin(x^3) \cdot \frac{d}{dx}(x^3)$$

$$\frac{dy}{dx} = -\sin(x^3) \cdot 3x^2$$

And because there was an $(x^3)$ in the original function, an $(x^3)$ must appear in the derivative.

## Example

Differentiate $h(x) = e^{5x}$.

## Solution

$$h(x) = e^{(\text{something})}$$

so:

$$h'(x) = e^{(\text{something})} \cdot \frac{d}{dx}(\text{something})$$

$$h'(x) = e^{5x} \cdot \frac{d}{dx}(5x) = e^{5x} \cdot 5 = 5e^{5x}$$

## ▶ Practice

Differentiate the following.

1. $f(x) = (8x^3 + 7)^4$

2. $y = (x^2 + 8x + 9)^3$

3. $h(t) = (t^8 - 9t^3 + 3t + 2)^{10}$

4. $y = (u^5 - 3u^4 + 7)^{\frac{7}{2}}$

5. $g(x) = \sqrt{x^2 + 9x + 1}$

6. $y = \sqrt[3]{e^x + 1}$

7. $f(x) = \sin(x^2)$

8. $g(x) = \sin^2(x)$

9. $y = \ln(3t + 5)$

10. $h(x) = \cos(3x)$

11. $y = e^{(x^2)}$

12. $y = \ln(x + 1)$

13. $s(u) = \cos^5(u)$

14. $y = (\ln(x))^5$

15. $f(x) = e^x + e^{2x} + e^{3x}$

16. $y = \tan(e^x)$

17. $g(x) = \dfrac{e^x - e^{-x}}{2}$

18. $f(\theta) = \dfrac{\sin(2\theta)}{\theta}$

19. $y = xe^{2x}$

20. $f(x) = \sec(10x^2 + e^x)$

This rule is called the Chain Rule because it works in long succession when there are many layers to the function. It helps to write out the function using lots of parentheses, and then work patiently to take the derivative of each outermost layer.

## Example
Differentiate $f(x) = \sin^7(e^{5x})$.

## Solution
With all of its parentheses, this function is $f(x) = (\sin(e^{(5x)}))^7$. The outermost layer is "something to the seventh power," the second layer is "the sine of something," the third layer is "$e$ raised to the something," and the last layer is $5x$. Thus:

$$f'(x) = 7(\sin(e^{(5x)}))^6 \cdot \frac{d}{dx}(\sin(e^{(5x)}))$$

$$f'(x) = 7(\sin(e^{(5x)}))^6 \cdot \cos(e^{(5x)}) \cdot \frac{d}{dx}(e^{(5x)})$$

$$f'(x) = 7(\sin(e^{(5x)}))^6 \cdot \cos(e^{(5x)}) \cdot e^{(5x)} \cdot \frac{d}{dx}(5x)$$

$$f'(x) = 7(\sin(e^{(5x)}))^6 \cdot \cos(e^{(5x)}) \cdot e^{(5x)} \cdot 5$$

$$f'(x) = 35e^{(5x)}\sin^6(e^{(5x)})\cos(e^{(5x)})$$

## Example
Differentiate $y = \ln(x^3 + \tan(3x^2 + x))$.

## Solution

$$\frac{dy}{dx} = \frac{1}{x^3 + \tan(3x^2 + x)} \cdot \frac{d}{dx}(x^3 + \tan(3x^2 + x))$$

$$\frac{dy}{dx} = \frac{1}{x^3 + \tan(3x^2 + x)} \cdot \left(3x^2 + \sec^2(3x^2 + x) \cdot \frac{d}{dx}(3x^2 + x)\right)$$

$$\frac{dy}{dx} = \frac{1}{x^3 + \tan(3x^2 + x)} \cdot (3x^2 + \sec^2(3x^2 + x) \cdot (6x + 1))$$

$$\frac{dy}{dx} = \frac{3x^2 + \sec^2(3x^2 + x) \cdot (6x + 1)}{x^3 + \tan(3x^2 + x)}$$

Notice once again that every part except the outermost layer (the natural logarithm) appears somewhere in the derivative.

## ▶ Practice

Differentiate the following.

**21.** $f(x) = \cos^3(8x)$

**22.** $y = (e^{9x^2 + 2x + 1})^4$

**23.** $g(t) = \ln(\tan(e^t + 1))$

**24.** $y = \sin(\sin(\sin(x)))$

**25.** $k(u) = \sec(\ln(8u^3))$

**26.** $h(x) = \ln(\cos(x + e^{3x}))$

# 11 ▶ Implicit Differentiation

A common complaint about the Chain Rule is "I don't know where to stop!" For example, why do we use the Chain Rule for $f(x) = \sin(x^3)$ to get $f'(x) = \cos(x^3) \cdot 3x^2$, but not for $g(x) = \sin(x)$, which has $g'(x) = \cos(x)$? The honest answer is that we *could* use the Chain Rule everywhere including in the following:

$$g'(x) = \cos(x) \cdot \frac{d}{dx}(x) = \cos(x) \cdot 1 = \cos(x)$$

$$f'(x) = \cos(x^3) \cdot \frac{d}{dx}(x^3) = \cos(x^3) \cdot 3x^2 \cdot \frac{d}{dx}(x) = \cos(x^3) \cdot 3x^2 \cdot 1 = \cos(x^3) \cdot 3x^2$$

When we get down to $\frac{d}{dx}(x) = 1$, we know we are done. The advantage to this way of thinking is that it explains what $\frac{dy}{dx}$ really means. This isn't merely a symbol that says "we took the derivative." This is the result of differentiating both sides of an equation.

## Example

Differentiate $y = 4x^5 + e^x$.

## Solution

Start with the equation.

$$y = 4x^5 + e^x$$

Differentiate both sides of the equation.

$$\frac{d}{dx}(y) = \frac{d}{dx}(4x^5 + e^x)$$

Use $\frac{d}{dx}(y) = \frac{dy}{dx}$.

$$\frac{dy}{dx} = 20x^4 \cdot \frac{d}{dx}(x) + e^x \cdot \frac{d}{dx}(x)$$

Simplify.

$$\frac{dy}{dx} = 20x^4 \cdot 1 + e^x \cdot 1 = 20x^4 + e^x$$

Now if $y = 4x^5 + e^x$, then there is a relationship between $y$ and $x$. This relationship is given *explicitly* because we know exactly what $y$ is in terms of $x$. However, if the variables $x$ and $y$ are all mixed up on both sides of the equals sign, then the relationship is given *implicitly*. The relationship is implied, but it is up to us to figure out what the relationship is explicitly. For example, the equation of the unit circle is:

$$x^2 + y^2 = 1$$

There is a relationship between the values of $x$ and $y$, because what $y$ can be depends on the value of $x$. If $x = 0$, for instance, then $y$ could be either 1 or $-1$. We could take the implicit description of $y$ in $x^2 + y^2 = 1$ and make it explicit by solving for $y$:

$$y^2 = 1 - x^2$$

$$y = \pm\sqrt{1 - x^2}$$

Solving for $y$ is not always possible, though. If our equation were $\ln(y) + \cos(y) = 3e^x - x^3$, then we would not be able to solve for $y$.

Fortunately, we can still find the slope of the tangent line, $\frac{dy}{dx}$, without having to solve the original equation for $y$. The trick is to use *implicit differentiation* by taking the derivative of both sides and making sure to include $\frac{d}{dx}(y) = \frac{dy}{dx}$ wherever the Chain Rule dictates.

## Example

Find the slope of the tangent line to $x^2 + y^2 = 1$.

## Solution

Start with the equation.

$$x^2 + y^2 = 1$$

Differentiate both sides.

$$\frac{d}{dx}(x^2 + y^2) = \frac{d}{dx}(1)$$

Use the Chain Rule everywhere.

$$2x \cdot \frac{d}{dx}(x) + 2y \cdot \frac{d}{dx}(y) = 0$$

Use $\frac{d}{dx}(x) = 1$ and $\frac{d}{dx}(y) = \frac{dy}{dx}$.

$$2x \cdot 1 + 2y \cdot \frac{dy}{dx} = 0$$

Solve for $\frac{dy}{dx}$.

$$\frac{dy}{dx} = \frac{-2x}{2y} = -\frac{x}{y}$$

It might feel unpleasant to have $\frac{dy}{dx}$ given in terms of both $x$ and $y$, but this is necessary. If we were asked, "What is the slope of the tangent line to $x^2 + y^2 = 1$ at $x = \dfrac{1}{2}$?" We would have to reply, "Which one?" There are *two* tangent lines with $x = \dfrac{1}{2}$! See Figure 11.1. If we want the slope of the tangent line at $\left( \dfrac{1}{2}, -\dfrac{\sqrt{3}}{2} \right)$, then

$$\frac{dy}{dx} = -\frac{x}{y} = -\frac{\frac{1}{2}}{-\frac{\sqrt{3}}{2}} = \frac{1}{\sqrt{3}} = \frac{\sqrt{3}}{3}.$$

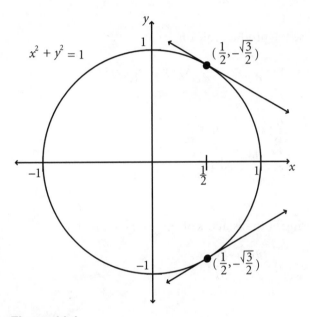

$x^2 + y^2 = 1$

$\left( \dfrac{1}{2}, -\dfrac{\sqrt{3}}{2} \right)$

$\left( \dfrac{1}{2}, -\dfrac{\sqrt{3}}{2} \right)$

**Figure 11.1**

## Example

Find $\dfrac{dy}{dx}$ when $\ln(y) + \cos(y) = 3e^x - x^3$.

## Solution

Start with the equation.

$\ln(y) + \cos(y) = 3e^x - x^3$

Differentiate both sides of the equation.

$\dfrac{d}{dx}(\ln(y) + \cos(y)) = \dfrac{d}{dx}(3e^x - x^3)$

Use the Chain Rule everywhere.

$$\frac{1}{y} \cdot \frac{d}{dx}(y) - \sin(y) \cdot \frac{d}{dx}(y) =$$

$$3e^x \cdot \frac{d}{dx}(x) - 3x^2 \cdot \frac{d}{dx}(x)$$

Use $\dfrac{d}{dx}(x) = 1$ and $\dfrac{d}{dx}(y) = \dfrac{dy}{dx}$.

$$\frac{1}{y} \cdot \frac{dy}{dx} - \sin(y) \cdot \frac{dy}{dx} = 3e^x - 3x^2$$

Factor out a $\dfrac{dy}{dx}$.

$$\left( \frac{1}{y} - \sin(y) \right) \frac{dy}{dx} = 3e^x - 3x^2$$

Solve for $\dfrac{dy}{dx}$.

$$\frac{dy}{dx} = \frac{3e^x - 3x^2}{\frac{1}{y} - \sin(y)}$$

To get rid of the fraction-in-a-fraction, we can multiply the top and bottom by the denominator $y$ that we want to eliminate:

$$\frac{dy}{dx} = \frac{3e^x - 3x^2}{\frac{1}{y} - \sin(y)}$$

$$= \left( \frac{3e^x - 3x^2}{\frac{1}{y} - \sin(y)} \right) \cdot \left( \frac{y}{y} \right)$$

$$= \frac{3ye^x - 3x^2y}{1 - y\sin(y)}$$

## Example

Find the slope of the tangent line to $y^2\ln(x) = y + 5$ at $(1, -5)$.

## Solution

Start with the equation.

$$y^2\ln(x) = y + 5$$

Differentiate both sides of the equation.

$$\frac{d}{dx}(y^2\ln(x)) = \frac{d}{dx}(y + 5)$$

Use the product rule on $y^2\ln(x)$.

$$2y \cdot \frac{d}{dx}(y) \cdot \ln(x) + \frac{1}{x} \cdot \frac{d}{dx}(x) \cdot y^2 = \frac{d}{dx}(y) + 0$$

Use $\frac{d}{dx}(x) = 1$ and $\frac{d}{dx}(y) = \frac{dy}{dx}$.

$$2y \cdot \frac{dy}{dx} \cdot \ln(x) + \frac{1}{x} \cdot y^2 = \frac{dy}{dx}$$

Plug in $x = 1$ and $y = -5$.

$$2(-5) \cdot \frac{dy}{dx} \cdot \ln(1) + \frac{1}{1} \cdot (-5)^2 = \frac{dy}{dx}$$

Use $\ln(1) = 0$.

$$25 = \frac{dy}{dx}$$

Thus, the slope of the tangent line at $(1, -5)$ is $\frac{dy}{dx} = 25$.

## Example

Find $\frac{dy}{dx}$ when $\tan(y) = xy + 7$.

## Solution

Start with the equation.

$$\tan(y) = xy + 7$$

Differentiate both sides of the equation.

$$\frac{d}{dx}(\tan(y)) = \frac{d}{dx}(xy + 7)$$

Use the product rule on $xy$.

$$\sec^2(y) \cdot \frac{d}{dx}(y) = \frac{d}{dx}(x) \cdot y + \frac{d}{dx}(y) \cdot x + 0$$

Use $\frac{d}{dx}(x) = 1$ and $\frac{d}{dx}(y) = \frac{dy}{dx}$.

$$\sec^2(y) \cdot \frac{dy}{dx} = y + \frac{dy}{dx} \cdot x$$

Bring both instances of $\frac{dy}{dx}$ to the same side.

$$\sec^2(y) \cdot \frac{dy}{dx} - \frac{dy}{dx} \cdot x = y$$

Factor out a $\frac{dy}{dx}$.

$$(\sec^2(y) - x)\frac{dy}{dx} = y$$

Solve for $\frac{dy}{dx}$.

$$\frac{dy}{dx} = \frac{y}{\sec^2(y) - x}$$

## Example

Now we are able to use implicit differentiation and the fact that $\frac{d}{dx}(e^x) = e^x$ to prove that $\frac{d}{dx}(\ln(x)) = \frac{1}{x}$.

## Solution

If $y = \ln(x)$, then the derivative will be $\frac{dy}{dx}$.

$y = \ln(x)$

Raise both sides as powers of $e$.

$e^y = e^{\ln(x)}$

Since $\ln(x)$ and $e^x$ are inverses, $e^{\ln(x)} = x$.

$e^y = x$

Differentiate both sides.

$\frac{d}{dx}(e^y) = \frac{d}{dx}(x)$

Use the Chain Rule.

$e^y \cdot \frac{dy}{dx} = 1$

Solve for $\frac{dy}{dx}$.

$\frac{dy}{dx} = \frac{1}{e^y}$

Use $e^y = e^{\ln(x)} = x$.

$\frac{dy}{dx} = \frac{1}{x}$

## ▶ Practice

Find $\frac{dy}{dx}$ in the following equations.

**1.** $(y + 1)^3 = x^4 - 8x$

**2.** $y^3 + y = \sin(x)$

**3.** $\sin(y) = 4x + 7$

**4.** $y - \sqrt{y} = \ln(x)$

**5.** $y^2 + x = 3x^4 + 8y$

**6.** $e^x + e^y = x^3$

**7.** $\tan(y) = \cos(x)$

**8.** $y = \sqrt{x + y}$

**9.** $\sin(x) - \sin(y) = x$

**10.** $y - \ln(y) = 10x^3 - 6x^2 + 4$

**11.** $(y + x^2)^4 = 10x$

**12.** $x^2y = y^4x^4$

**13.** $\frac{x}{y} + xy = x + y$

**14.** $\sec(y) + 9y = x^3\cos(y)$

**15.** Find the tangent line slope of
$y^3 + x^2 = y^2 - 5y + 14$ at $(-3,1)$.

**16.** Find the tangent line slope of $x^3 + y^3 = 3y - x$ at $(1,-2)$.

**17.** Find the slope of the tangent line to
$\ln(3y - 5) + x = y^2$ at (4,2).

**18.** Find the slope of the tangent line at (2,3) on the graph of $x^2 y + y^2 x = 30$.

**19.** Find the equation of the tangent line to $\sin(y) = x$ at the point $\left(\dfrac{1}{2}, \dfrac{\pi}{6}\right)$.

**20.** Find the equation of the tangent line to $x^2 + 6y = xy + 3$ at (3,−2).

L E S S O N

# 12 ▶ Related Rates

Once you have gotten the hang of implicit differentiation, it should not be difficult to take the derivative of both sides with respect to the variable $t$. This enables us to see how $x$ and $y$ vary with respect to time $t$. The only difference is that $\frac{d}{dt}(x) = \frac{dx}{dt}$, $\frac{d}{dt}(y) = \frac{dy}{dt}$, and so on. Only $\frac{d}{dt}(t) = 1$ can be simplified, but this generally never occurs.

## Example
Differentiate $y^2 + \cos(x) = 4x^2y$ with respect to $t$.

## Solution
Start with the equation.

$y^2 + \cos(x) = 4x^2y$

Differentiate both sides with respect to $t$.

$\frac{d}{dt}(y^2 + \cos(x)) = \frac{d}{dt}(4x^2y)$

Use the Chain Rule everywhere.

$$2y \cdot \frac{d}{dt}(y) - \sin(x) \cdot \frac{d}{dt}(x) =$$

$$8x \cdot \frac{d}{dt}(x) \cdot y + \frac{d}{dt}(y) \cdot 4x^2$$

Use $\frac{d}{dt}(x) = \frac{dx}{dt}$ and $\frac{d}{dt}(y) = \frac{dy}{dt}$.

$$2y \cdot \frac{dy}{dt} - \sin(x) \cdot \frac{dx}{dt} = 8xy \cdot \frac{dx}{dt} + \frac{dy}{dt} \cdot 4x^2$$

## Example

Differentiate $e^x + y = y^3 + \sqrt{x}$ with respect to $t$.

## Solution

Start with the equation.

$$e^x + y = y^3 + \sqrt{x}$$

Differentiate both sides with respect to $t$.

$$\frac{d}{dt}(e^x + y) = \frac{d}{dt}(y^3 + \sqrt{x})$$

Use the Chain Rule everywhere.

$$e^x \cdot \frac{d}{dt}(x) + \frac{d}{dt}(y) = 3y^2 \cdot \frac{d}{dt}(y) + \frac{1}{2\sqrt{x}} \cdot \frac{d}{dt}(x)$$

Use $\frac{d}{dt}(x) = \frac{dx}{dt}$ and $\frac{d}{dt}(y) = \frac{dy}{dt}$.

$$e^x \cdot \frac{dx}{dt} + \frac{dy}{dt} = 3y^2 \cdot \frac{dy}{dt} + \frac{1}{2\sqrt{x}} \cdot \frac{dx}{dt}$$

The variables need not be $x$ and $y$.

## Example

Differentiate $3A + 4B^2 = \frac{A}{r}$ with respect to $t$.

## Solution

$$\frac{d}{dt}(3A + 4B^2) = \frac{d}{dt}\left(\frac{A}{r}\right)$$

$$3 \cdot \frac{dA}{dt} + 8B \cdot \frac{dB}{dt} = \frac{\frac{dA}{dt} \cdot r - \frac{dr}{dt} \cdot A}{r^2}$$

## Example

Differentiate $A = \pi r^2$ with respect to $t$.

## Solution

$$\frac{d}{dt}(A) = \frac{d}{dt}(\pi r^2)$$

$$\frac{dA}{dt} = 2\pi r \cdot \frac{dr}{dt}$$

Don't forget that $\pi$ is a constant, not a variable!

## ▶ Practice

Differentiate with respect to $t$.

**1.** $y = (x^3 + x - 1)^5$

**2.** $y^4 - 3x^2 = \cos(y)$

**3.** $y^3 - y = 3x^4 - 10x^2 + 3x + 1$

**4.** $\sqrt{x} + \sqrt{y} = 10x^3 - 7x$

**5.** $\ln(y) + e^x = x^2y^2$

**6.** $5x^2 + 2x + 1 = w^2 + 7$

**7.** $z = \frac{2}{5}x^2 + \frac{2}{5}y^2 + \frac{3}{5x}$

**8.** $A^2 + B^2 = C^2$

**9.** $V = \frac{4}{3}\pi r^3$

**10.** $A = 4\pi r^2$

**11.** $C = 2\pi r$

**12.** $A = \frac{1}{2}bh$

Just as $\frac{dy}{dx} = \frac{y\text{-change}}{x\text{-change}}$ is a rate, so are $\frac{dx}{dt}$, $\frac{dy}{dt}$, $\frac{dA}{dt}$, and so on. Because $t$ usually represents time, $\frac{dy}{dt} = \frac{y\text{-change}}{t\text{-change}}$ represents how fast $y$ is changing over time. Thus, if $A$ is a variable that represents an area, $\frac{dA}{dt}$ represents how fast that area is increasing or decreasing.

Differentiating an equation with respect to $t$ results in a new equation, which shows how the rates of change of the variables are related. For example, the area and radius of a circle are related by:

$A = \pi r^2$

If we differentiate with respect to $t$, we get:

$\frac{dA}{dt} = 2\pi r \cdot \frac{dr}{dt}$

If a circle is growing in size, this equation details how the rate at which the radius is changing, $\frac{dr}{dt}$, relates to the rate at which the area is growing, $\frac{dA}{dt}$.

## Example

A rock thrown into a pond makes a circular ripple that travels at 4 feet per second. How fast is the area of the circle increasing when the circle has a radius of 12 feet?

## Solution

We know that for circles, $\frac{dA}{dt} = 2\pi r \cdot \frac{dr}{dt}$. And we know that the radius is increasing at the rate of $\frac{dr}{dt} = 4$ feet per second, so when the radius is $r = 12$ feet, the area is increasing at:

$\frac{dA}{dt} = 2\pi(12 \text{ feet}) \cdot 4\frac{\text{feet}}{\text{second}}$

$= 96\pi \frac{\text{ft}^2}{\text{sec}}$

$= 96\pi \approx 301.6$ square feet per second

## Example

A spherical balloon is inflated with 40 cubic inches of air every second. When the radius is 12 inches, how fast is the radius of the balloon increasing? (Hint: The volume of a sphere with radius $r$ is $V = \frac{4}{3}\pi r^3$.)

## Solution

We know that the volume of the balloon is increasing at the rate of $\frac{dV}{dt} = 40\frac{\text{in}^3}{\text{sec}}$. We want to know what $\frac{dr}{dt}$ is when $r = 12$ inches. If we differentiate $V = \frac{4}{3}\pi r^3$ with respect to $t$, we get:

$\frac{dV}{dt} = 4\pi r^2 \cdot \frac{dr}{dt}$

When we plug in $\frac{dV}{dt} = 40\frac{\text{in}^3}{\text{sec}}$ and $r = 12$ in, we get:

$40\frac{\text{in}^3}{\text{sec}} = 4\pi(12\text{ in})^2 \cdot \frac{dr}{dt}$

$\frac{dr}{dt} = \frac{40 \text{ in}}{4\pi \cdot 144 \text{ sec}} = \frac{5\text{in}}{72\pi\text{sec}}$ $\frac{40}{576}$

The radius of the balloon is increasing at the very slow rate of $\frac{5}{72\pi} \approx 0.022$ inches per second.

## Example

Suppose the base of a triangle is increasing at a rate of 8 feet per minute while the height is decreasing by 1 foot every minute. How fast is the triangle's area changing when the height is 5 feet and the base is 20 feet?

## Solution

If we represent the length of the base by $b$, the height of the triangle as $h$, and the area of the triangle as $A$, then the formula that relates them all is $A = \frac{1}{2}bh$. The base is increasing at $\frac{db}{dt} = 8\frac{\text{ft}}{\text{min}}$ and the height is changing at $\frac{dh}{dt} = -1\frac{\text{ft}}{\text{min}}$. The $-1$ implies that 1 foot is subtracted from the height every minute, that is, the height is decreasing. We are trying to find $\frac{dA}{dt}$, which is the rate of change in area. When we differentiate our formula $A = \frac{1}{2}bh$ with respect to $t$, we get:

$$\frac{dA}{dt} = \frac{1}{2} \cdot \frac{db}{dt} \cdot h + \frac{dh}{dt} \cdot \frac{1}{2}b$$

When we plug in all of our information, including the $h = 5$ feet and $b = 20$ feet, we get:

$$\frac{dA}{dt} = \frac{1}{2} \cdot (8) \cdot (5) + (-1) \cdot \frac{1}{2} \cdot (20)$$
$$= 20 - 10 = 10$$

Thus, at the exact instant when the height is 5 feet and the base is 20, the area of the triangle is increasing at a rate of 10 square feet every minute.

## Example

A 20 foot ladder slides down a wall at the rate of 2 feet per minute (see Figure 12.1). How fast is it sliding along the ground when the ladder is 16 feet up the wall?

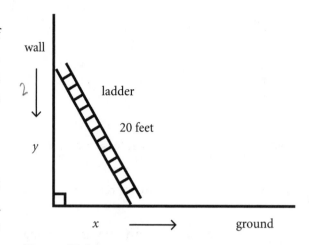

**Figure 12.1**

## Solution

Here, $\frac{dy}{dt} = -2\frac{\text{ft}}{\text{min}}$ because the ladder is sliding down the wall at 2 feet per minute. We want to know $\frac{dx}{dt}$, the rate at which the bottom of the ladder is moving away from the wall. The equation to use is the Pythagorean theorem.

$$x^2 + y^2 = 20^2$$

$$\frac{d}{dt}(x^2 + y^2) = \frac{d}{dt}(20^2)$$

$$2x \cdot \frac{dx}{dt} + 2y \cdot \frac{dy}{dt} = 0$$

If we plug in $y = 16$ and $\frac{dy}{dt} = -2$, we get:

$$2x \cdot \frac{dx}{dt} + 2(16) \cdot (-2) = 0$$

We still need to know what $x$ is at the particular instant that $y = 16$, and for this, we go back to the Pythagorean theorem.

$$x^2 + (16)^2 = (20)^2$$

$$x^2 = 144, \text{ so } x = \pm 12$$

It is important to use variables for all of the values that are changing. Only after differentiation can they be replaced by numbers.

---

Using $x = 12$ (a negative length here makes no sense), we get:

$$2 \cdot (12) \cdot \frac{dx}{dt} + 2 \cdot (16)(-2) = 0$$

$$\frac{dx}{dt} = \frac{8}{3}$$

At the moment that $y = 16$, the ladder is sliding along the ground at $\frac{8}{3}$ feet per minute.

In the previous example, it was okay to say that the hypotenuse was 20 because the length of the ladder didn't change. However, if we replace $y$ with 16 in the equation before differentiating, we would have implied that the height was fixed at 16 feet. Because the height does change, it needs to be written as a variable, $y$. In general, anything that varies needs to be represented with a variable. Only after the derivative has been taken can the information for the given instant, like $y = 16$, be substituted.

### ▶ Practice

**13.** Suppose $y^2 + 3y = 6 - 4x^3$ and $\frac{dy}{dt} = 5$. What is $\frac{dx}{dt}$ when $x = -1$ and $y = 2$?

**14.** Suppose $xy^2 = x^2 + 3$. What is $\frac{dy}{dt}$ when $\frac{dx}{dt} = 8$, $x = 3$, and $y = -2$?

**15.** Let $K + e^L = L + I^2$. If $\frac{dL}{dt} = 5$ and $\frac{dI}{dt} = 4$, what is $\frac{dK}{dt}$ when $L = 0$ and $I = 3$?

**16.** Suppose $A^3 = B^2 + 4C^2$, $\frac{dA}{dt} = 8$, and $\frac{dC}{dt} = -2$. What is $\frac{dB}{dt}$ when $A = 2$, $B = 2$, and $C = 1$?

**17.** Suppose $A = I^2 + 6R$. If $I$ increases by 4 feet per minute and $R$ increases by 2 square feet every minute, how fast is $A$ changing when $I = 20$?

**18.** Suppose $K^3 = \frac{1}{R^2} + 11$. Every hour, $K$ increases by 2. How fast is $R$ changing when $K = 3$ and $R = \frac{1}{4}$?

**19.** The height of a triangle increases by 2 feet every minute while its base shrinks by 6 feet every minute. How fast is the area changing when the height is 15 feet and the base is 20 feet?

**20.** The surface area of a sphere with radius $r$ is $A = 4\pi r^2$. If the radius is decreasing by 2 inches every minute, how fast is the surface area shrinking when the radius is 20 inches?

**21.** A circle increases in area by 20 square feet every hour. How fast is the radius increasing when the radius is 4 feet?

**22.** The volume of a cube grows by 1,200 square inches every minute. How fast is each side growing when each side is 10 inches?

**23.** The height of a triangle grows by 5 inches each hour. The area is increasing by 100 square inches each hour. How fast is the base of the triangle increasing when the height is 20 inches and the base is 12 inches?

**24.** One end of a 10-foot long board is lifted straight off the ground at 1 foot per second (see figure below). How fast will the other end drag along the ground after 6 seconds?

**25.** A kite is 100 feet off the ground and moving horizontally at 13 feet per second (see figure below). How quickly must the string be let out when the string is 260 feet long?

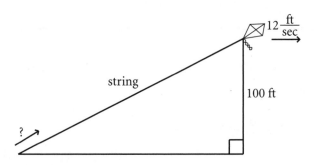

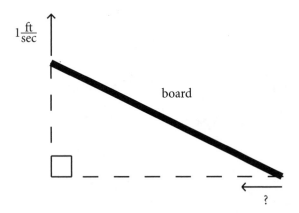

# 13 ▶ Limits at Infinity

This lesson will serve as a preparation for the graphing in the next lesson. Here, we will work on ways to identify asymptotes from the formula of a rational function. *Rational functions* are quotients, with a clear numerator and denominator.

*Vertical asymptotes* are easy to recognize, because they occurwhere the denominator is undefined. For example, $f(x) = \dfrac{(3x + 2)(x - 1)}{(x + 3)(x - 4)}$ has vertical asymptotes at $x = -3$ and $x = 4$.

*Horizontal asymptotes* take a bit more work to identify. The graph will flatten out like a horizontal line if large values of $x$ all have essentially the same $y$-value.

In this graph of $y = f(x)$, for example, if $x$ is bigger than 5, then $y$ will be very close to $y = 1$ (see Figure 13.1). Thus, $y = 1$ is a horizontal asymptote. Similarly, if $x$ is a large negative number, the corresponding $y$-value will be close to zero. Thus, $y = 0$ is another horizontal asymptote. Horizontal asymptotes are related to the limits as $x$ gets really big. For $f(x)$ given in the graph:

$$\lim_{x\to\infty} f(x) = 1 \ \text{ and } \ \lim_{x\to -\infty} f(x) = 0$$

Notice that the graph of $y = f(x)$ crosses *both* asymptotes. Vertical asymptotes cannot be crossed because they are, by definition, not in the domain. Horizontal asymptotes *can* be crossed, as illustrated in this example. Think of "asymptote" as meaning "flattens out like a straight line" and not "a line not to be crossed."

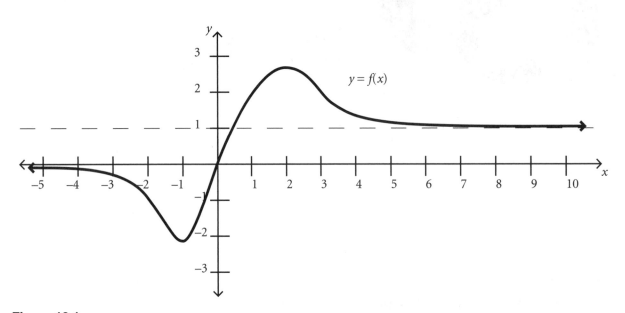

**Figure 13.1**

These *limits at infinity* (and negative infinity) identify what the ends of the graph do. For example, if $\lim_{x \to \infty} g(x) = 3$, then the graph of $y = g(x)$ will look something like that in Figure 13.2. If $\lim_{x \to -\infty} h(x) = \infty$, then the graph of $y = h(x)$ will look like that in Figure 13.3.

Notice that the infinite limits say only what happens way off to the left and to the right. Other calculations must be done to know what happens in the middle of the graph.

The general trick to evaluating an infinite limit is to focus on the most powerful part of the function. Take $\lim_{x \to \infty} 2x^3 - 100x^2 - 10x - 5,000$, for example.

There are a lot of negative elements to this function. However, the most powerful part is the positive $2x^3$. When $x$ gets big enough, like when $x = 1,000,000$, then

$$2x^3 - 100x^2 - 10x - 5,000$$

$$= 2,000,000,000,000,000,000 -$$

$$100,000,000,000,000 - 10,000,000 - 5,000$$

$$= 1,999,899,999,989,995,000$$

This clearly rounds to 2,000,000,000, 000,000,000, which is the $2x^3$. It is in this sense that

The rules for Infinite Limits of Rational Functions are as follows:
- If the numerator is more powerful, the limit goes to ∞ or −∞.
- If the denominator is more powerful, the limit goes to 0.
- If the numerator and denominator are evenly matched, the limit is formed by the coefficients of the most powerful parts.

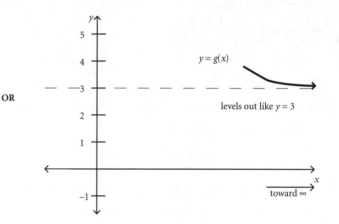
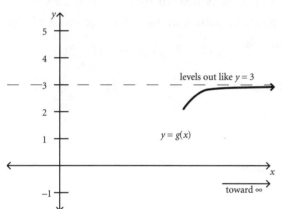

**Figure 13.2**

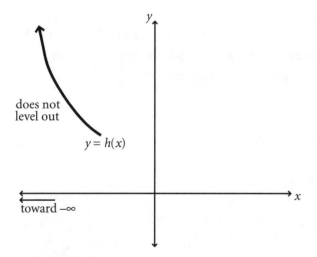

**Figure 13.3**

$2x^3$ is called the most powerful part of the function. As $x$ gets big, $2x^3$ is the only part that counts.

$$\lim_{x \to \infty} 2x^3 - 100x^2 - 10x - 5{,}000 = \lim_{x \to \infty} 2x^3 = \infty$$

As $x$ gets huge, $x^3$ is clearly even larger, and $2x^3$ is twice that. Thus, as $x$ goes to infinity, so does $2x^3$. Basically, the higher the exponent of $x$, the more powerful it is. With that in mind, the rules for infinite limits of rational functions are fairly simple:

- If the numerator is more powerful, the limit goes to ∞ or −∞.
- If the denominator is more powerful, the limit goes to 0.
- If the numerator and denominator are evenly matched, the limit is formed by the coefficients of the most powerful parts.

The whole concept of "going to infinity" might be a bit confusing. This really means "going toward infinity," because infinity is not something that a real number can reach. So don't wander off pondering the one number that is bigger than all the rest (unless you enjoy that). Just know that "going to infinity" means using really big numbers, and that "going to negative infinity" means using really big negative numbers.

## Example

Evaluate $\lim\limits_{x \to \infty} \dfrac{1 - x^2}{x^3 + 3x + 2}$.

## Solution

The most powerful part of the numerator is $-x^2$, and in the denominator is $x^3$. Thus:

$$\lim_{x \to \infty} \frac{1 - x^2}{x^3 + 3x + 2} = \lim_{x \to \infty} \frac{-x^2}{x^3} = \lim_{x \to \infty} -\frac{1}{x} = 0$$

This goes to zero because the numerator is clearly outclassed by the more powerful denominator. Also, as $x$ gets really big, $\dfrac{1}{x}$ gets really close to zero. For example, when $x = 1{,}000$, then $\dfrac{1}{x} = \dfrac{1}{1{,}000} = 0.001$.

## Example

Evaluate $\lim\limits_{x \to -\infty} \dfrac{3x^2 + 2x - 5}{1 - 8x^2}$.

## Solution

Here, the numerator and denominator are evenly matched, with each having $x^2$ as its highest power of $x$.

$$\lim_{x \to -\infty} \frac{3x^2 + 2x - 5}{1 - 8x^2} = \lim_{x \to -\infty} \frac{3x^2}{-8x^2}$$
$$= \lim_{x \to -\infty} -\frac{3}{8} = -\frac{3}{8}$$

The limit is formed by the coefficients of the most powerful parts: 3 in the numerator and $-8$ in the denominator.

## Example

Evaluate $\lim\limits_{x \to \infty} \dfrac{5x^{10} - 4x^5 + 7}{1 - x^2}$.

## Solution

Here,

$$\lim_{x \to \infty} \frac{5x^{10} - 4x^5 + 7}{1 - x^2} = \lim_{x \to \infty} \frac{5x^{10}}{-x^2}$$
$$= \lim_{x \to \infty} -5x^8 = -\infty$$

As $x$ goes to infinity, $x^8$ also gets really large, but the negative in the $-5$ reverses this and makes $-5x^8$ approach negative infinity.

## ▶ Practice

Evaluate the following infinite limits.

**1.** $\lim\limits_{x \to \infty} \dfrac{5x^3 + 10x^2 - 2}{8x^4 + 1}$

**2.** $\lim\limits_{x \to -\infty} \dfrac{4x^3 + 10x^2 + 3x}{5x^3 + 8x - 1}$

**3.** $\lim\limits_{x \to \infty} \dfrac{5x + 2}{2x - 1}$

**4.** $\lim\limits_{x\to\infty} \dfrac{10x^3 - 3x - 100}{2x + 5}$

**5.** $\lim\limits_{t\to -\infty} \dfrac{t + 1}{t^3 + 3t - 4}$

**6.** $\lim\limits_{t\to\infty} \dfrac{8t^4 - 3t^3 + 11}{1 - 9t^4}$

**7.** $\lim\limits_{x\to\infty} \dfrac{5x^3 + x - 9}{1 - x^2}$

**8.** $\lim\limits_{x\to -\infty} \dfrac{x^4 + 3x^2 - 8x + 4}{x^2 + 2x + 1}$

**9.** $\lim\limits_{x\to -\infty} \dfrac{x^2 - 1}{x^2 + 1}$

**10.** $\lim\limits_{x\to\infty} \dfrac{t^{10} + 4t^2 - 11}{2t^{15} - 7t}$

The infinite limits of $e^x$ and $\ln(x)$ can be seen from their graphs in Figure 13.4.

$$\lim\limits_{x\to\infty} e^x = \infty \qquad \lim\limits_{x\to -\infty} e^x = 0 \qquad \lim\limits_{x\to\infty} \ln(x) = \infty$$

In general, as $x$ goes to infinity, $e^x$ is more powerful than $x$ raised to any number. The natural logarithm, however, goes to infinity slower than just about anything else. It may look as though $y = \ln(x)$ is beginning to level out into a horizontal asymptote, but actually, it will eventually surpass any height as it slowly goes up to infinity.

In more complicated situations, we use L'Hôpital's rule. This states that if the numerator and denominator both go to infinity (positive or negative), then the limit remains the same after taking the derivative of the top and the bottom.

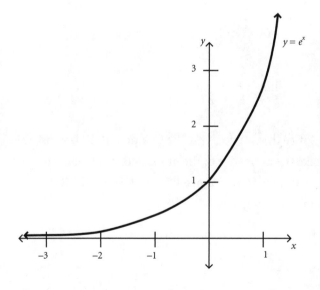

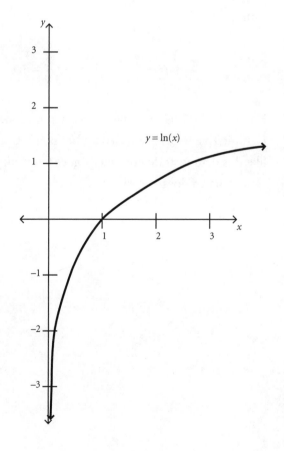

**Figure 13.4**

# L'Hôpital's Rule

If the numerator and denominator both go to infinity (positive or negative), the limit remains the same after taking the derivative of the top and bottom, OR:

$$\lim_{x \to \pm\infty} \frac{f(x)}{g(x)} = \lim_{x \to \pm\infty} \frac{f'(x)}{g'(x)} \text{ if } \lim_{x \to \pm\infty} f(x) = \pm\infty \text{ and } \lim_{x \to \pm\infty} g(x) = \pm\infty$$

## Example

Evaluate $\lim\limits_{x \to \infty} \dfrac{\ln(x)}{1 - x}$.

## Solution

Since $\lim\limits_{x \to \infty} \ln(x) = \infty$ and $\lim\limits_{x \to \infty} 1 - x = -\infty$, we can use L'Hôpital's Rule.

$$\lim_{x \to \infty} \frac{\ln(x)}{1 - x} \overset{H}{=} \lim_{x \to \infty} \frac{\frac{d}{dx}(\ln(x))}{\frac{d}{dx}(1 - x)}$$

$$= \lim_{x \to \infty} \frac{\frac{1}{x}}{-1} = \lim_{x \to \infty} -\frac{1}{x} = 0$$

**Note:** The little $H$ over the equals sign indicates that L'Hôpital's Rule as been used at that point. Examples like this demonstrate how $\ln(x)$ goes to infinity even slower than $x$ does.

## Example

Evaluate $\lim\limits_{x \to \infty} \dfrac{e^x}{x^3 + 2x^2 + 5x + 2}$.

## Solution

Here, $\lim\limits_{x \to \infty} e^x = \infty$ and $\lim\limits_{x \to \infty} x^3 + 2x^2 + 5x + 2 = \infty$, so we use L'Hôpital's Rule.

$$\lim_{x \to \infty} \frac{e^x}{x^3 + 2x^2 + 5x + 2}$$

$$\overset{H}{=} \lim_{x \to \infty} \frac{\frac{d}{dx}(e^x)}{\frac{d}{dx}(x^3 + 2x^2 + 5x + 2)}$$

$$= \lim_{x \to \infty} \frac{e^x}{3x^2 + 4x + 5}$$

Notice that we don't use the Quotient Rule, because we take the derivative of the numerator and denominator separately. Here, we need to use L'Hôpital's Rule several more times:

$$\lim_{x \to \infty} \frac{e^x}{3x^2 + 4x + 5} \overset{H}{=} \lim_{x \to \infty} \frac{e^x}{6x + 4}$$

$$\overset{H}{=} \lim_{x \to \infty} \frac{e^x 6}{} = \infty$$

This example shows how $e^x$ is more powerful than $x^3$. If the denominator had an $x^{100}$, we'd have to use L'Hôpital's Rule 100 times, but in the end, the $e^x$ would take everything to infinity.

## Example

Evaluate $\lim\limits_{x \to -\infty} \dfrac{e^x}{x^5 + 7x - 1}$.

## Solution

This is a trick question! The limit $\lim\limits_{x \to -\infty} e^x = 0$ is not infinite, so we *can't* use L'Hôpital's Rule. The function $e^x$ is only powerful when $x$ goes to positive infinity. Instead, we use the old "plug in" method (or common sense).

$$\lim\limits_{x \to -\infty} \dfrac{e^x}{x^5 + 7x - 1} = \dfrac{0}{\text{something not zero}} = 0$$

## Example

Evaluate $\lim\limits_{x \to \infty} \dfrac{\sin(x)}{x^2}$.

## Solution

This has the same problem as the previous example. No matter what $x$ may be, $\sin(x)$ will always be between $-1$ and $1$. Thus, $-1 \le \sin(x) \le 1$ and so

$$\dfrac{-1}{x^2} \le \dfrac{\sin(x)}{x^2} \le \dfrac{1}{x^2}$$

Because $\lim\limits_{x \to \infty} \dfrac{1}{x} = 0$ and $\lim\limits_{x \to \infty} \dfrac{-1}{x} = 0$, the function $\dfrac{\sin(x)}{x^2}$ is squeezed between them to zero as well: $\lim\limits_{x \to \infty} \dfrac{\sin(x)}{x^2} = 0$. This is called the *Squeeze Theorem* or the *Sandwich Theorem* because of the way $\dfrac{\sin(x)}{x^2}$ is squished between something above it going to zero and something below it going to zero.

## ▶ Practice

Evaluate the following limits.

11. $\lim\limits_{x \to \infty} \dfrac{\ln(x^3)}{\ln(x) + 5}$

12. $\lim\limits_{x \to \infty} \dfrac{x + 5}{\sqrt{x - 1}}$

13. $\lim\limits_{x \to -\infty} \dfrac{x^2 + 5x - 10}{4x + 2}$

14. $\lim\limits_{x \to \infty} \dfrac{3x^2 + 2}{x - \ln(x)}$

15. $\lim\limits_{x \to -\infty} \dfrac{4x^3 - 10x + 7}{15x^3 + 4x^2 - 2x + 1}$

16. $\lim\limits_{x \to \infty} \dfrac{4x + 6}{3x^2 - 2x + 5}$

17. $\lim\limits_{x \to \infty} \dfrac{3x + 7}{\ln(x)}$

18. $\lim\limits_{x \to \infty} \dfrac{\cos(x)}{x}$

19. $\lim\limits_{x \to \infty} \dfrac{4x^3 + 5x^2 + 2}{e^x - 7x^3}$

20. $\lim\limits_{x \to -\infty} \dfrac{4x^3 + 5x^2 + 2}{e^x - 7x^3}$

## ▶ Sign Diagrams

In order to calculate the limits at vertical asymptotes, it is necessary to know where the function is positive and negative. The key to everything is this: A continuous function cannot switch between positive and negative without being zero or undefined. Functions are generally zero when the numerator is zero and undefined where the denominator is zero. Mark these points on a number line. Between these points, the function must be entirely positive or negative. This can be found by testing any point in each interval.

For example, consider $f(x) = \dfrac{x - 4}{(x + 2)(1 - x)}$. This function is zero at $x = 4$ and undefined at both $x = -2$ and $x = 1$. We mark these on a number line (see Figure 13.5).

In between $x = -2$ and $x = 1$, the function is either always positive or always negative. To find out which it is, we test a point between $-2$ and $1$, such as 0. Because $f(0) = \dfrac{-4}{2(1)} = -2$ is negative, the function is always negative between $-2$ and 1. Similarly, we check a point between 1 and 4, such as $f(2) = \dfrac{-2}{4(-1)} = \dfrac{1}{2}$, a point after 4, such as $f(5) = \dfrac{1}{7(-4)} = -\dfrac{1}{28}$, and a point before $-2$, such as $f(-3) = \dfrac{-7}{-1(4)} = \dfrac{7}{4}$. These calculations can be made very roughly, because it matters only if the function is positive or negative at the selected point. In any case, the *sign diagram* for this function is shown in Figure 13.6.

This makes calculating the limits at the vertical asymptotes very easy. Not only does

$f(x) = \dfrac{x - 4}{(x + 2)(1 - x)}$ have vertical asymptotes at $x = -2$ and $x = 1$, but the limits are:

$$\lim_{x \to -2^-} \frac{x - 4}{(x + 2)(1 - x)} = \infty$$

$$\lim_{x \to -2^+} \frac{x - 4}{(x + 2)(1 - x)} = -\infty$$

$$\lim_{x \to 1^-} \frac{x - 4}{(x + 2)(1 - x)} = -\infty$$

$$\lim_{x \to 1^+} \frac{x - 4}{(x + 2)(1 - x)} = \infty$$

At the same time, we can calculate the limits at infinity:

$$\lim_{x \to \infty} \frac{x - 4}{(x + 2)(1 - x)} = \lim_{x \to \infty} \frac{x - 4}{-x^2 - x + 2} = 0$$

$$\lim_{x \to -\infty} \frac{x - 4}{(x + 2)(1 - x)} = 0$$

Thus, $f(x)$ has a horizontal asymptote of $y = 0$. With all of this, we begin to get a picture of $f(x) = \dfrac{x - 4}{(x + 2)(1 - x)}$, which can be seen in Figure 13.7.

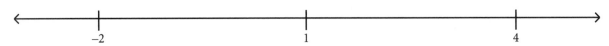

**Figure 13.5**

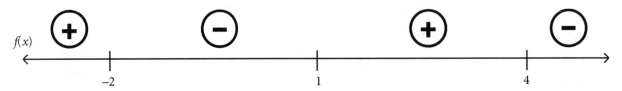

**Figure 13.6**

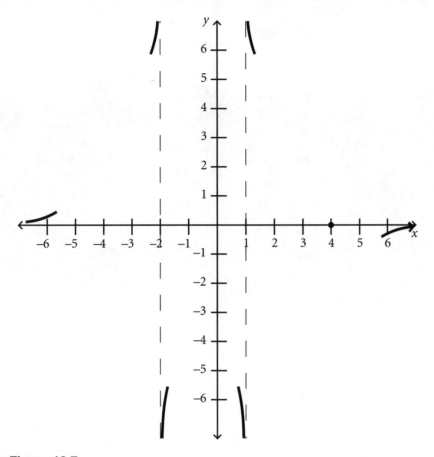

**Figure 13.7**

Notice that the horizontal asymptote $y = 0$ is approached from above as $x \to -\infty$, because $f(x)$ is always positive when $x < -2$. At the other end, the asymptote is approached from below as $x \to \infty$ because the function is negative when $x > 4$.

We shall deal with graphing more thoroughly in the next lesson.

### ▶ Practice

Name all the asymptotes, vertical and horizontal, of the following functions. Also, make a sign diagram for each.

**21.** $f(x) = \dfrac{x + 2}{x - 4}$

**22.** $g(x) = \dfrac{x - 3}{x^2 - 4}$

**23.** $h(x) = \dfrac{x^2 - 1}{(x + 3)^2}$

**24.** $k(x) = \dfrac{2x + 1}{x^2 - 4x + 3}$

Evaluate the following limits.

**25.** $\displaystyle\lim_{x \to 4^+} \dfrac{x + 2}{x - 4}$

**26.** $\displaystyle\lim_{x \to 2^-} \dfrac{x - 3}{x^2 - 4}$

**27.** $\displaystyle\lim_{x \to -3^-} \dfrac{x^2 - 1}{(x + 3)^2}$

**28.** $\displaystyle\lim_{x \to 3^-} \dfrac{x + 1}{x^2 - 4x + 3}$

# 14 ▶ Graphs

Here is where everything comes together. We know how to find the domain, how to identify asymptotes, and how to plot points. With the help of the sign diagrams from the previous lesson, we shall be able to tell where a function is increasing and decreasing, and where it is concave up and down.

Quite simply, where the derivative is positive, the function is increasing. The derivative gives the slope of the tangent line at a point, and when this is positive, the function is heading upward, viewed from left to right. When the derivative is negative, the function slopes downward and decreases.

When the second derivative is positive, the function is concave up. This is because the second derivative says how the first derivative is changing. If the second derivative is positive, then the slopes are increasing. If the slopes, from left to right, increase from 22, to 21, to 0, to 1, to 2, and so on, then the graph must curve like the one in Figure 14.1. In other words, the curve must be concave up.

Similarly, if the second derivative is negative, the function curves downward like the one in Figure 14.2 and is concave down.

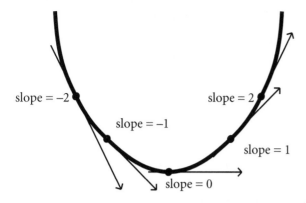

Figure 14.1

Figure 14.2

The concavity governs the shape of the graph, depending on whether the function $f(x)$ is increasing or decreasing. If $f(x)$ is increasing and concave up (thus, both $f'(x)$ and $f''(x)$ are positive), then the graph has the shape shown in Figure 14.3.

increasing    +    concave up    =

Figure 14.3

If $f(x)$ is increasing and concave down (thus, $f'(x)$ is positive and $f''(x)$ is negative), then the graph has the shape shown in Figure 14.4.

increasing    +    concave down    =

Figure 14.4

If $f(x)$ is decreasing and concave down (thus, both $f'(x)$ and $f''(x)$ are negative), then the graph has the shape of the one in Figure 14.5.

decreasing    +    concave down    =

Figure 14.5

If $f(x)$ is decreasing and concave up (thus, $f'(x) < 0$ and $f''(x) > 0$), the graph has the shape of the one in Figure 14.6.

decreasing    +    concave up    =

Figure 14.6

## Example

Graph $f(x) = x^3 + 6x^2 - 15x + 10$.

## Solution

This function is defined everywhere and thus has no vertical asymptotes. Because $\lim_{x \to \infty} x^3 + 6x^2 - 15x + 10 = \infty$ and $\lim_{x \to -\infty} x^3 + 6x^2 - 15x + 10 = -\infty$, there are no horizontal asymptotes.

The derivative $f'(x) = 3x^2 + 12x - 15 = 3(x^2 + 4x - 5) = 3(x + 5)(x - 1)$ is zero at $x = -5$ and $x = 1$. To form the sign diagram, we test: $f'(-6) = 21$, $f'(0) = 15$, and $f'(2) = 21$. **Note:** These points were chosen arbitrarily. Any point less than $-5$ will give the same information as the value $x = -6$, for instance, and any point between $-5$ and $-1$ will give the same information as the value at $x = 0$. Thus, the sign diagram for $f'(x)$ is shown in Figure 14.7.

Remember, the sign of $f'(x)$ determines whether $f(x)$ is increasing or decreasing.

**Note:** We use $f'(x)$ to see if the graph is increasing or decreasing, but $f(x)$ to find the $y$-value at a point.

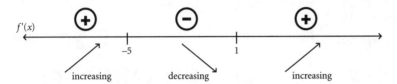

**Figure 14.7**

Because the function increases up to $x = -5$ and then decreases immediately afterward, there is a local maximum at $x = -5$. The corresponding $y$-value is $y = f(-5) = 110$. Thus, $(-5,110)$ is a local maximum. Similarly, the graph goes down to $x = 1$ and then goes up afterward, so $x = 1$ is a local minimum. The corresponding $y$-value is $f(1) = 2$, so $(1,2)$ is a local minimum. $f'(x)$ and $f(x)$

A guideline for identifying local minimum and maximum points is shown in Figure 14.8.

The second derivative is $f''(x) = 6x + 12 = 6(x + 2)$, which is zero at $x = -2$. If we test the sign at $x = -3$ and $x = 0$, we get $f''(-3) = -6$ and $f''(0) = 12$. Thus, the sign diagram for $f''(x)$ is as shown in Figure 14.9.

**Figure 14.8**

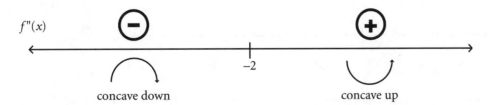

**Figure 14.9**

Clearly $x = -2$ is a point of inflection, because this is where the concavity switches from concave down to concave up. The $y$-value of this point is $f(-2) = 56$.

Before we draw the axes for the Cartesian plane, we should consider the three interesting points we have found: the local maximum at $(-5, 110)$, the local minimum at $(1, 2)$, and the point of inflection at $(-2, 56)$. If our $x$-axis runs from $x = -10$ to $x = 10$, and our $y$-axis runs from 0 to 120, then all of these can be plotted easily (see Figure 14.10).

## Example

Graph $g(x) = \dfrac{x + 3}{x - 2}$.

## Solution

The domain is $x \neq 2$. There is a vertical asymptote at $x = 2$. The sign diagram or $g(x)$ is shown in Figure 14.11.

Thus, $\displaystyle\lim_{x \to 2^-} \dfrac{x + 3}{x - 2} = -\infty$ and $\displaystyle\lim_{x \to 2^+} \dfrac{x + 3}{x - 2} = \infty$.

Because $\displaystyle\lim_{x \to \infty} \dfrac{x + 3}{x - 2} = 1$ and $\displaystyle\lim_{x \to -\infty} \dfrac{x + 3}{x - 2} = 1$, there

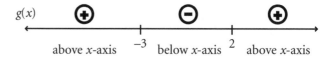

**Figure 14.11**

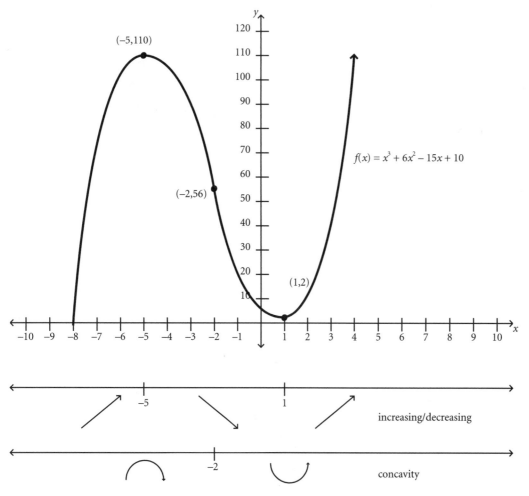

$f(x) = x^3 + 6x^2 - 15x + 10$

**Figure 14.10**

is a horizontal asymptote at $y = 1$, both to the left and to the right. The derivative $g'(x) = \dfrac{1 \cdot (x - 2) - 1 \cdot (x + 3)}{(x - 2)^2} = \dfrac{-5}{(x - 2)^2}$ has the sign diagram shown in Figure 14.12.

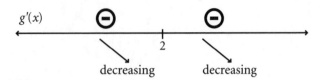

**Figure 14.12**

The second derivative $g''(x) = \dfrac{10}{(x - 2)^3}$ has sign diagram shown in Figure 14.13.

Because we have no points plotted at all, it makes sense to pick one or two to the left and right of the ver-

**Figure 14.13**

tical asymptote at $x = 2$. At $x = 1$, $g(1) = -4$, so $(1, -4)$ is a point. At $x = 3$, $g(3) = 6$, so $(3, 6)$ is another point. At $x = -3$, $g(-3) = 0$, so $(-3, 0)$ is another nice point to know. Judging by these, it will be useful to have both the $x$- and $y$-axes run from $-10$ to 10.

To graph $g(x)$, it helps to start with the points and the asymptotes as shown in Figure 14.14.

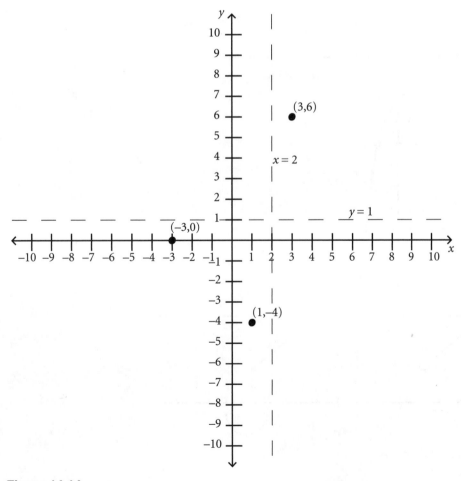

**Figure 14.14**

Then we establish the shapes of the lines through these points using the concavity and the intervals of decrease (see Figure 14.15).

## Example

Graph $h(x) = \dfrac{x^2 + 1}{x^2 - 1}$.

## Solution

To start, $h(x) = \dfrac{x^2 + 1}{x^2 - 1} = \dfrac{x^2 + 1}{(x + 1)(x - 1)}$. Thus, $h(x)$ is undefined with a vertical asymptote at $x = 1$ and $x = -1$. The sign diagram for $h(x)$ is shown in Figure 14.16.

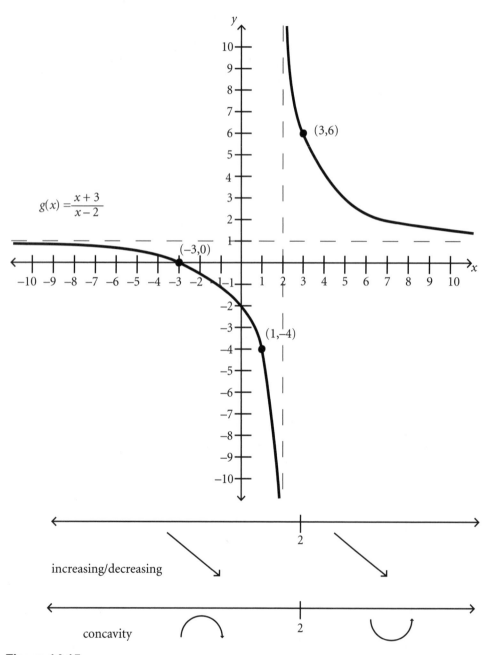

increasing/decreasing

concavity

**Figure 14.15**

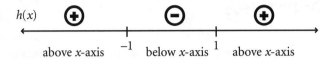

above $x$-axis    below $x$-axis    above $x$-axis

**Figure 14.16**

**Note:** $x^2 + 1$ can never be zero. The limits at the vertical asymptotes are thus:

$$\lim_{x \to -1^-} \frac{x^2 + 1}{x^2 - 1} = \infty$$

$$\lim_{x \to -1^+} \frac{x^2 + 1}{x^2 - 1} = -\infty$$

$$\lim_{x \to 1^-} \frac{x^2 + 1}{x^2 - 1} = -\infty$$

$$\lim_{x \to 1^+} \frac{x^2 + 1}{x^2 - 1} = \infty$$

Because $\lim\limits_{x \to \infty} \dfrac{x^2 + 1}{x^2 - 1} = 1$ and $\lim\limits_{x \to -\infty} \dfrac{x^2 + 1}{x^2 - 1} = 1$, there is a horizontal asymptote at $y = 1$.

The derivative is as follows:

$$h'(x) = \frac{2x(x^2 - 1) - 2x(x^2 1)}{(x^2 - 1)^2} = \frac{-4x}{(x - 1)^2(x + 1)^2}$$

It has the sign diagram shown in Figure 14.17. This indicates that there is a local maximum at $x = 0$. The corresponding $y$-value is $y = h(0) = -1$.

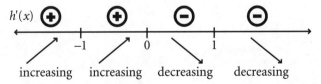

increasing    increasing    decreasing    decreasing

**Figure 14.17**

The second derivative is as follows:

$$h''(x) = \frac{-4(x^2 - 1)^2 - 2(x^2 - 1) \cdot 2x(-4x)}{(x^2 - 1)^4}$$

$$= \frac{-4(x^2 - 1) - 2 \cdot 2x(-4x)}{(x^2 - 1)^3}$$

$$= \frac{12x^2 + 4}{(x^2 - 1)^3} = \frac{12x^2 + 4}{(x - 1)^3(x + 1)^3}.$$

The sign diagram is shown in Figure 14.18. It looks like there ought to be points of inflection at $x = -1$ and $x = 1$, but these are asymptotes not in the domain, so there are no points where the concavity changes.

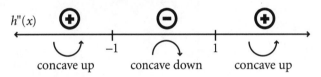

concave up    concave down    concave up

**Figure 14.18**

Before we graph the function, it will be useful to have a few more points. When $x = -2$, then $y = h(-2) = \dfrac{5}{3}$ and when $x = 2$, $y = h(2) = \dfrac{5}{3}$ as well. Thus, it will be useful to have the $x$- and $y$-axes run from $-3$ to $3$. We start with just the points and asymptotes (see Figure 14.19).

Then we add in the actual curves, guided by the concavity and the intervals of increase and decrease (see Figure 14.20).

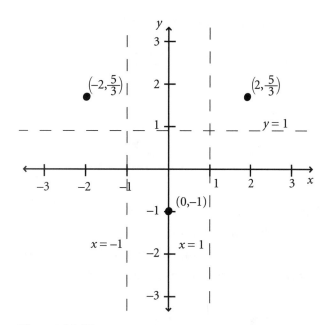

Figure 14.19

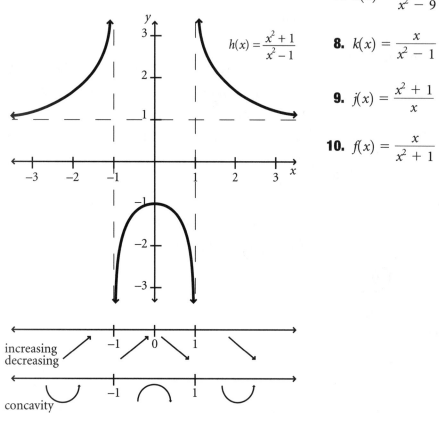

Figure 14.20

## ▶ Practice

Use the asymptotes, concavity, and intervals of increase and decrease to graph the following functions.

**1.** $f(x) = x^2 - 30x + 10$

**2.** $g(x) = -4x - x^2$

**3.** $h(x) = 2x^3 - 3x^2 - 36x + 5$

**4.** $k(x) = 3x - x^3$

**5.** $f(x) = x^4 - 8x^3 + 5$

**6.** $g(x) = \dfrac{x}{x + 2}$

**7.** $h(x) = \dfrac{1}{x^2 - 9}$

**8.** $k(x) = \dfrac{x}{x^2 - 1}$

**9.** $j(x) = \dfrac{x^2 + 1}{x}$

**10.** $f(x) = \dfrac{x}{x^2 + 1}$

# 15 ▶ Optimization

Knowing the minimum and maximum points of a function is useful for graphing and even more useful in real-life situations. Businesses want to maximize their profits, builders want to minimize their costs, drivers want to minimize distances, and people want to get the most for their money. If we can represent a situation with a function, then the derivative will help find the optimal point.

If the derivative is zero or undefined at exactly one point, then this is very likely to be the optimal point. The *first derivative test* states that if the function increases before that point and decreases afterward, it is maximal (see Figure 15.1). Similarly, if the function decreases before the point and increases afterward, then the point is an absolute minimum.

The *second derivative test* states that if the second derivative is positive, then the function curves up, so a point of slope zero must be a minimum (see Figure 15.2). Similarly, if the second derivative is negative, the point of slope zero must be the highest point on the graph. Remember that we are assuming that only one point has slope zero or an undefined derivative.

If there are several points of slope zero and the function has a closed interval for a domain, then plug all the *critical points* (points of slope zero, points of undefined derivative, and the two endpoints of the interval) into the original function. The point with the highest $y$-value will be the absolute maximum, and the one with the smallest $y$-value will be the absolute minimum.

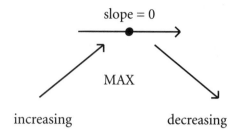

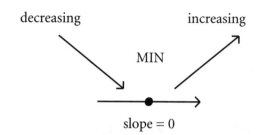

slope = 0 (MAX) increasing decreasing

decreasing increasing MIN slope = 0

**Figure 15.1**

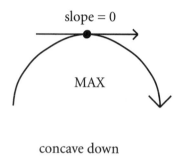

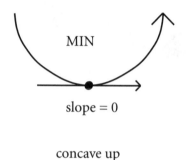

slope = 0 MAX concave down

MIN slope = 0 concave up

**Figure 15.2**

## Example

A manager calculates that when $x$ employees are working at the same time, the store makes a profit of $P(x) = 15x^2 - 48x - x^3$ dollars each hour. If there are ten employees and at least one must be working at any given time, how many employees should be scheduled to maximize profit?

## Solution

This is an instance of a function on a closed interval because $1 \le x \le 10$ limits the options for $x$. The derivative of the profit function is $P'(x) = 30x - 48 - 3x^2$ which factors into $P'(x) = -3(x^2 - 10x + 16) = -3(x - 2)(x - 8)$. Thus, the derivative is zero at $x = 2$ and at $x = 8$.

Because more than two points have a slope of zero, we cannot use the first or second derivative tests. Instead, we evaluate each of our critical points. These are the points of slope zero, $x = 2$ and $x = 8$, plus the

endpoints of our interval $x = 1$ and $x = 10$. These are evaluated as follows: $P(1) = -34$, $P(2) = -44$, $P(8) = 64$, and $P(10) = 20$. If the manager wants to maximize the store profit, eight employees should be scheduled at the time, because this will result in a maximal profit of $64 each hour.

## Example

A coffee shop owner calculates that if she sells cookies at $p$ each, she will sell $\dfrac{200}{p^2}$ cookies each day. If it costs her 20¢ to make each cookie, what price $p$ will give her the greatest profit?

## Solution

The function for profit is Profit = Revenue − Costs. If she charges $p$ per cookie, then she'll make and sell $\dfrac{200}{p^2}$ cookies each day. Thus, her revenue will be

$\left(\dfrac{200}{p^2}\right) \cdot p = \dfrac{200}{p}$ and her costs will be $\left(\dfrac{200}{p^2}\right) \cdot (0.20)$. Therefore, her profit function is $\text{Profit}(p) = \dfrac{200}{p} - \dfrac{40}{p^2}$. We limit this to $p > 0.20$ because the only optimal situation would be when the cookies were sold for more than it cost to make them.

The derivative is $\text{Profit}'(p) = -\dfrac{200}{p^2} + \dfrac{80}{p^3}$, which is zero when $\dfrac{80}{p^3} = \dfrac{200}{p^2}$ and therefore $80p^2 = 200p^3$, so either $p = 0$ or else $p = \dfrac{80}{200} = 0.40$. Because $p = 0$ is not in the domain, the only place where the derivative is zero is at $p = 40¢$.

Using the first derivative test, we see that $\text{Profit}'(0.30) = 740$ and $\text{Profit}'(0.50) = -160$, therefore our sign diagram for Profit' is as shown in Figure 15.3. So the absolute maximal profit occurs when the cookies are sold at $40¢$.

## Example

At $1 per cup of coffee, a vendor sells 500 cups a day. When the price is increased to $1.10, the vendor sells only 480 cups. If every 1¢ increase in price reduces the sales by two cups, what price per cup of coffee will maximize income?

## Solution

Here, the income is Income = Price $\times$ Cups Sold. So if $x =$ the number of pennies the price is increased, then $\text{Income}(x) = (1 + 0.01x) \cdot (500 - 2x)$. This simplifies to $\text{Income}(x) = 500 + 3x - 0.02x^2$. And, the derivative is $\text{Income}'(x) = 3 - 0.04x$. This is zero only when $x = \dfrac{3}{0.04} = 75$. The second derivative is $\text{Income}''(x) = -0.04$, which is negative, so $x = 75$ is maximal by the second derivative test. Thus, the maximal income will occur when the price is raised by $x = 75¢$ to $1.75 per cup.

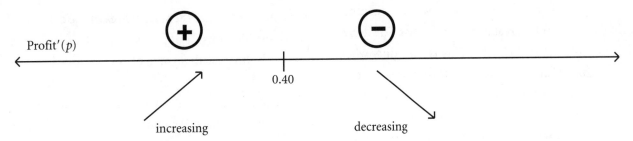

Profit'$(p)$

$+$      $-$

0.40

increasing      decreasing

**Figure 15.3**

## Example

A farmer wants to build a rectangular pen with 80 feet of fencing. The pen will be built against the side of a barn, so one side won't need a fence. What dimensions will maximize the area of the pen? See Figure 15.4.

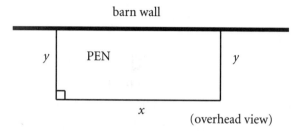

barn wall

$y$ | PEN | $y$

$x$

(overhead view)

**Figure 15.4**

## Solution

The area of the pen is Area $= x \cdot y$. We can't take the derivative of this just yet because there are two variables. We need to use the additional information regarding how much fencing exists; there are 80 feet of fencing. Because no fencing will be required against the barn wall, the total lengths of the fence will be $y + x + y = 80$, thus $x = 80 - 2y$. We can plug this into the formula for area in order to obtain Area $= x \cdot y = (80 - 2y) \cdot y$. Now we have a function of one variable Area$(y) = 80y - 2y^2$. The derivative is Area$'(y) = 80 - 4y$. This is zero only when $y = 20$. Using the second derivative test, Area$''(y) = -4$, thus the curve is concave down and the point $y = 20$ is the absolute maximum. The corresponding $x$-value is $x = 80 - 2y = 80 - 2(20) = 40$. Therefore, the pen with the maximal area will be $x = 40$ feet wide (along the barn) and $y = 20$ feet out from the barn wall.

## Example

A manufacturer needs to design a crate with a square bottom and no top. It must hold exactly 32 cubic feet of shredded paper. What dimensions will minimize the material needed to make the crate (the surface area)? See Figure 15.5.

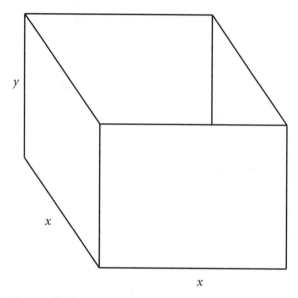

$y$

$x$

$x$

**Figure 15.5**

## Solution

We want to minimize the surface area of the crate. The area of the box consists of four sides, each of area $x \cdot y$, plus the bottom, with an area of $x \cdot x = x^2$. Thus, the surface area is Area $= 4xy + x^2$. Again, we need to reduce this to a formula with only one variable in order to differentiate. We know that the volume must be 32 cubic feet, so Volume $= x^2 y = 32$. Thus, $y = \dfrac{32}{x^2}$. When we plug this into the surface area function, we get:

$$\text{Area} = 4xy + x^2 = 4x\left(\frac{32}{x^2}\right) + x^2 = \frac{128}{x} + x^2.$$

The derivative is:

$$\text{Area}'(x) = -\frac{128}{x^2} + 2x,$$

which is zero when

$$-\frac{128}{x^2} + 2x = 0 \text{ or } x^3 = 64, \text{ so } x = 4.$$

The second derivative is:

$$\text{Area}''(x) = \frac{256}{x^3} + 2,$$

which is positive when $x = 4$. So the curve is concave up and the sole point of slope zero is the absolute minimum. Thus, the surface area of the crate will be minimized if $x = 4$ feet and $y = \frac{32}{x^2} = \frac{32}{4^2} = 2$ feet.

## ▶ Practice

1. Suppose a company makes a profit of
   $P(x) = \frac{1,000}{x} - \frac{5,000}{x^2} + 100$ dollars when it
   makes and sells $x > 0$ items. How many items
   should it make to maximize profit?

2. Suppose the profit of a company is
   $P(x) = 9x^2 + 40x - \frac{1}{3}x^3 + 1,000$ when it
   makes $x$ items a day. What level of production
   will maximize profits?

3. When 30 orange trees are planted on an acre,
   each will produce 500 oranges a year. For every
   additional orange tree planted, each tree will
   produce 10 fewer oranges. How many trees
   should be planted to maximize the yield?

4. An artist can sell 20 copies of a painting at $100
   each, but for each additional copy she makes, the
   value of each painting will go down by a dollar.
   Thus, if 22 copies are made, each will sell for $98.
   How many copies should she make to maximize
   her sales?

5. A garden has 200 pounds of watermelons growing in it. Every day, the total amount of watermelon increases by 5 pounds. At the same time, the price per pound of watermelon goes down by 1¢. If the current price is 90¢ per pound, how much longer should the watermelons grow in order to fetch the highest price possible?

6. A farmer has 400 feet of fencing to make three rectangular pens. What dimensions $x$ and $y$ will maximize the total area?

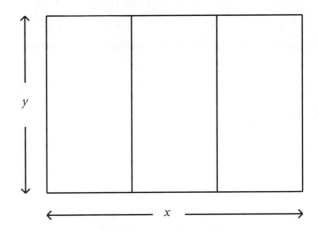

7. Four pens will be built along a river by using 150 feet of fencing. What dimensions will maximize the area of the pens?

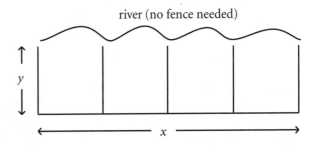

8. A rectangular pen will be built using 100 feet of fencing. What dimensions will maximize the area?

**9.** The surface area of a can is Area $=$ $2\pi r^2 + 2\pi rh$, where the height is $h$ and the radius is $r$. The volume is Volume $= \pi r^2 h$. What dimensions minimize the surface area of a can with volume $16\pi$ cubic inches?

**10.** A painter has enough paint to cover 600 square feet of area. What is the largest square-bottom box that could be painted (including the top, bottom, and all sides)?

**11.** A box with a square bottom will be built to contain 40,000 cubic feet of grain. The sides of the box cost 10¢ per square foot to build, the roof costs $1 per square foot to build, and the bottom will cost $7 per square foot to build. What dimensions will minimize the building costs?

**12.** A printed page will have a total area of 96 square inches. The top and bottom margins will be 1 inch each, and the left and right margins will be $1\frac{1}{2}$ inches each. What overall dimensions for the page will maximize the area of the space inside the margins?

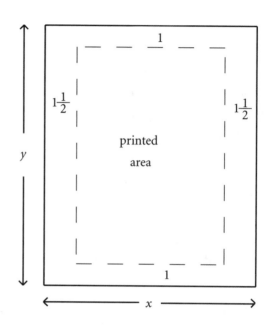

# L E S S O N

# 16 ▶ Areas under Curves

A round the same time that so many great mathematicians devoted themselves to figuring out the slopes of tangent lines, other mathematicians were working on an entirely different problem. They wanted to be able to figure out the area underneath any curve $y = f(x)$, such as the one shown in Figure 16.1.

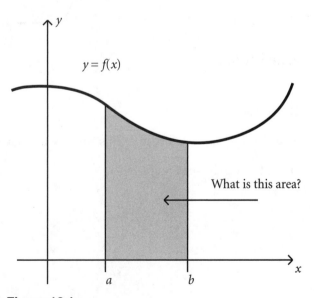

**Figure 16.1**

They used the shorthand notation $\int_a^b f(x)\,dx$ to represent the area between, or bound by, the curve $y = f(x)$, the x-axis, and the lines $x = a$ and $x = b$. This symbol is generally referred to as an *integral*. Note: The $dx$ is merely to indicate which letter is the variable. If the horizontal axis represented time, for instance, then this would end in $dt$.

## Example

Evaluate the integral $\int_0^4 \frac{1}{2}x\,dx$.

## Solution

This represents the area between the curve $y = \frac{1}{2}x$, the x-axis, the line $x = 0$, and the line $x = 4$ (see Figure 16.2). This area happens to be a triangle with a height of 2 and a base of 4. The area of the triangle is $\frac{1}{2}(2)(4) = 4$. Thus, $\int_0^4 \frac{1}{2}x\,dx = 4$.

Also, just to be sporting, area *below* the x-axis is counted negatively. Therefore, really $\int_a^b f(x)\,dx$ represents "the area between the curve $y = f(x)$, the x-axis, $x = a$, and $x = b$, where area below the x-axis is counted negatively."

## Example

Evaluate the integrals $\int_1^3 f(x)\,dx$, $\int_1^4 f(x)\,dx$, $\int_1^6 f(x)\,dx$, and $\int_6^7 f(x)\,dx$ where the graph of $y = f(x)$ is given as shown in Figure 16.3.

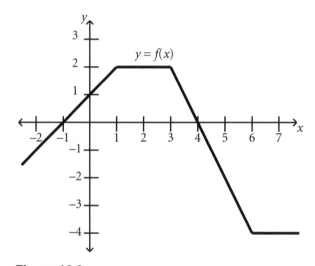

**Figure 16.3**

## Solution

First, $\int_1^3 f(x)\,dx = 4$ because this area is a square above the x-axis (see Figure 16.4).

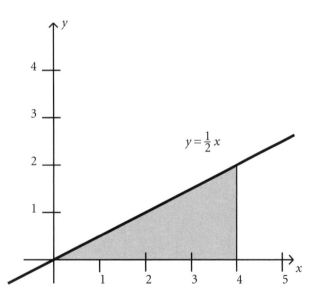

**Figure 16.2**

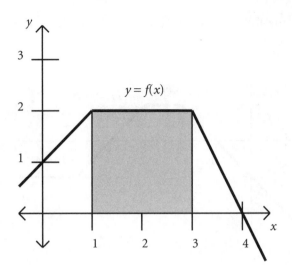

**Figure 16.4**

Next, $\displaystyle\int_{1}^{4} f(x)\,dx = 4 + \frac{1}{2} \cdot (1) \cdot 2 = 5$ because this area is a square plus a triangle (see Figure 16.5).

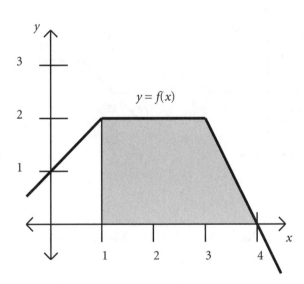

**Figure 16.5**

For $\displaystyle\int_{1}^{6} f(x)\,dx$, we must calculate how much area is above the $x$-axis and how much is below (see Figure 16.6).

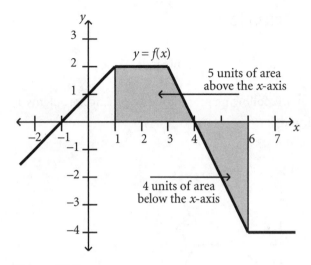

**Figure 16.6**

There are 5 units of area above the $x$-axis and 4 units below, so $\displaystyle\int_{1}^{6} f(x)\,dx = 5 - 4 = 1$.

Finally, $\displaystyle\int_{6}^{7} f(x)\,dx$ represents a rectangle of area 4 that is entirely below the $x$-axis. Thus, $\displaystyle\int_{6}^{7} f(x)\,dx = -4$ (see Figure 16.7).

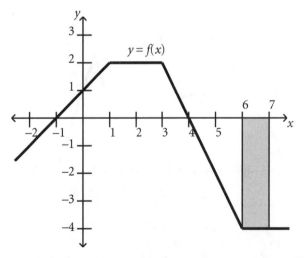

**Figure 16.7**

▶ **Practice**

Evaluate the following integrals.

Use the following graph to solve practice problems 1, 2, and 3.

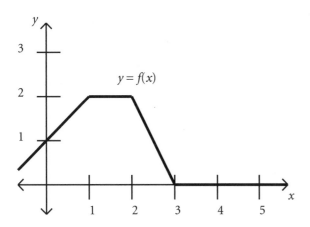

**1.** $\int_{1}^{2} f(x)\,dx$    **2.** $\int_{0}^{1} f(x)\,dx$    **3.** $\int_{0}^{2} f(x)\,dx$

Use the following graph for practice problems 4, 5, and 6.

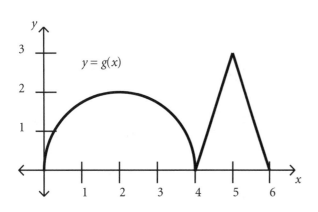

**4.** $\int_{0}^{4} g(x)\,dx$    **5.** $\int_{4}^{6} g(x)\,dx$    **6.** $\int_{0}^{6} g(x)\,dx$

Use the following graph for practice problems 7, 8, and 9.

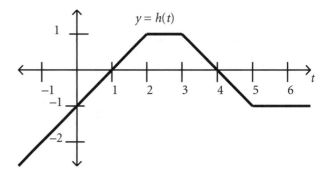

**7.** $\int_{-1}^{6} h(t)\,dt$    **8.** $\int_{-1}^{4} h(t)\,dt$    **9.** $\int_{4}^{6} h(t)\,dt$

Use the following graph for practice problems 10 through 18.

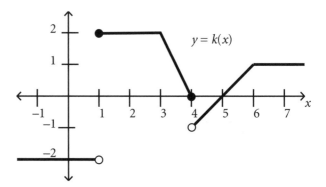

**10.** $\int_{0}^{7} k(x)\,dx$    **11.** $\int_{4}^{6} k(x)\,dx$    **12.** $\int_{4}^{5} k(x)\,dx$

**13.** $\int_{0}^{4} (x + 2)\,dx$ **14.** $\int_{1}^{4} 2\,dx$    **15.** $\int_{1}^{5} (t - 3)\,dt$

**16.** $\int_{-2}^{5} x\,dx$    **17.** $\int_{1}^{6} 2x\,dx$    **18.** $\int_{0}^{8} (2x - 2)\,dx$

You may have noticed:

$$\int_a^b f(x)\,dx + \int_b^c f(x)\,dx = \int_a^c f(x)\,dx$$

The area between $a$ and $c$ is the area from $a$ to $b$ plus the area from $b$ to $c$, assuming, of course, that $a < b < c$ (see Figure 16.8).

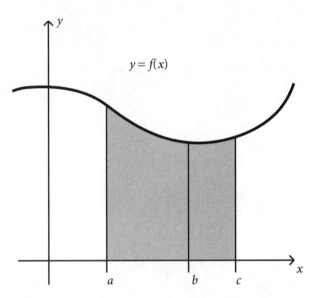

**Figure 16.8**

Similarly, $\int_a^c f(x)\,dx - \int_b^c f(x)\,dx = \int_a^b f(x)\,dx$.

We can use these to make calculations, even when the exact functions are unknown.

## Example

If $\int_3^5 f(x)\,dx = 7$ and $\int_5^{10} f(x)\,dx = 15$, then what is $\int_3^{10} f(x)\,dx$?

## Solution

$$\int_3^{10} f(x)\,dx = \int_3^5 f(x)\,dx + \int_5^{10} f(x)\,dx$$
$$= 7 + 15 = 22$$

## Example

If $\int_0^{10} g(x)\,dx = 38$ and $\int_8^{10} g(x)\,dx = -12$, then what is $\int_0^8 g(x)\,dx$?

## Solution

$$\int_0^8 g(x)\,dx = \int_0^{10} g(x)\,dx - \int_8^{10} g(x)\,dx$$
$$= 38 - (-12) = 50$$

## ▶ Practice

Suppose $\displaystyle\int_0^6 f(x)\,dx = 10$, $\displaystyle\int_6^7 f(x)\,dx = -5$, and $\displaystyle\int_7^{11} f(x)\,dx = 2$. Evaluate the following.

**19.** $\displaystyle\int_0^7 f(x)\,dx$    **20.** $\displaystyle\int_6^{11} f(x)\,dx$    **21.** $\displaystyle\int_0^{11} f(x)\,dx$

Suppose $\displaystyle\int_1^{14} g(t)\,dt = -3$, $\displaystyle\int_{10}^{14} g(t)\,dt = 8$, and $\displaystyle\int_1^5 g(t)\,dt = -10$. Evaluate the following.

**22.** $\displaystyle\int_5^{14} g(t)\,dt$    **23.** $\displaystyle\int_1^{10} g(t)\,dt$    **24.** $\displaystyle\int_5^{10} g(t)\,dt$

Suppose $\displaystyle\int_{-2}^{11} h(x)\,dx = 20$, $\displaystyle\int_{-2}^1 h(x)\,dx = 12$, and $\displaystyle\int_{-2}^{10} h(x)\,dx = -5$. Evaluate the following.

**25.** $\displaystyle\int_1^{11} h(x)\,dx$    **26.** $\displaystyle\int_1^{10} h(x)\,dx$    **27.** $\displaystyle\int_{10}^{11} h(x)\,dx$

# 17 ▶ The Fundamental Theorem of Calculus

**H**ere comes the resounding climax of calculus. It would be best to read this lesson with some bombastic orchestral music like that of Wagner or Orff. This, however, is not necessary. The initial question here is innocent enough: If we make a function from that "area under a curve" stuff, what would its derivative be? So suppose that our curve is $y = f(t)$ (see Figure 17.1). We use the variable $t$ in order to save $x$ for something more important.

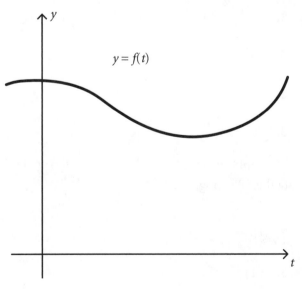

$y = f(t)$

**Figure 17.1**

Now let our "area under the curve function" be $g(x)$ = the area under the curve $y = f(t)$ between 0 and some point $x$. Therefore, $g(x) = \int_0^x f(t)\,dt$. This area is illustrated in Figure 17.2.

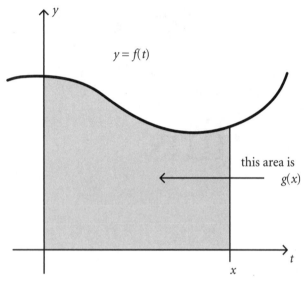

Figure 17.2

### Example
If $f(t) = 2t$ and $g(x) = \int_0^x f(t)\,dt$, then what is $g(3)$?

### Solution
$g(3) = \int_0^3 f(t)\,dt = \int_0^3 2t\,dt$ = the area beneath the curve $y = 2t$ from 0 to 3. The graph of $f(t) = 2t$ is shown in Figure 17.3. This area is a triangle with base 3 and height 6, so $g(3) = \int_0^3 2t\,dt = \frac{1}{2}(3)(6) = 9$.

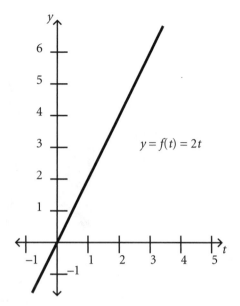

Figure 17.3

### ▶ Practice

Suppose $f(t) = \frac{1}{2}t + 1$ and $g(x) = \int_0^x f(t)\,dt$. Evaluate the following.

1. $g(1)$
2. $g(2)$
3. $g(3)$
4. $g(4)$
5. $g(5)$
6. $g(0)$

Now suppose $f(t) = 7$ and $g(x) = \int_0^x f(t)\,dt$. Evaluate the following.

7. $g(0)$
8. $g(1)$

**9.** $g(2)$

**10.** $g(3)$

**11.** $g(4)$

**12.** $g(5)$

Now that the concept of $g(x) = \displaystyle\int_0^x f(t)\,dt$ is clear, we can answer the next question: What is the derivative of $g(x)$?

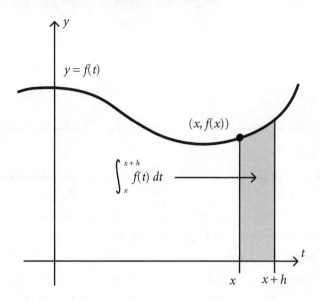

$y = f(t)$

$(x, f(x))$

$\displaystyle\int_x^{x+h} f(t)\,dt$

$x \qquad x+h$

**Figure 17.4**

Begin with the definition of the derivative.

$$g'(x) = \lim_{h \to 0} \frac{g(x+h) - g(x)}{h}$$

Use $g(x) = \displaystyle\int_0^x f(t)\,dt$.

$$g'(x) = \lim_{h \to 0} \frac{\displaystyle\int_0^{x+h} f(t)\,dt - \int_0^x f(t)\,dt}{h}$$

Use $\displaystyle\int_a^c f(t)\,dt - \int_a^b f(t)\,dt = \int_b^c f(t)\,dt$.

$$g'(x) = \lim_{h \to 0} \frac{\displaystyle\int_x^{x+h} f(t)\,dt}{h}$$

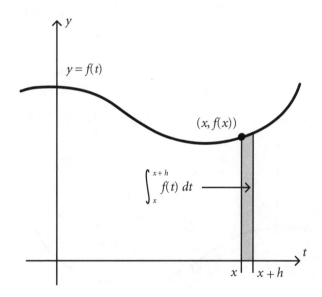

$y = f(t)$

$(x, f(x))$

$\displaystyle\int_x^{x+h} f(t)\,dt$

$x \quad x+h$

**Figure 17.5**

Now the integral $\displaystyle\int_x^{x+h} f(t)\,dt$ represents the skinny little area just to the right of point $x$ (see Figure 17.4). This is *almost* a rectangle with a base of $h$ and a height of $f(x)$, so the integral $\displaystyle\int_x^{x+h} f(t)\,dt$ is *almost* $h \cdot f(x)$. As $h$ gets really small, this area gets closer to being a rectangle (see Figure 17.5).

Therefore, as $h$ approaches zero, this integral approaches $h \cdot f(x)$, thus:

$$g'(x) = \lim_{h \to 0} \frac{\displaystyle\int_x^{x+h} f(t)\,dt}{h}$$

$$= \lim_{h \to 0} \frac{h \cdot f(x)}{h} = \lim_{h \to 0} f(x) = f(x)$$

The Fundamental Theorem of Calculus can be written as follows:

$$\int_a^b f(x)\,dx = \int_0^b f(x)\,dx - \int_0^a f(x)\,dx = g(b) - g(a)\,,\ \text{where}\ g'(x) = f(x)$$

What does this mean? It means that the derivative of the function $g(x)$, which represents "the area under the curve," is the very function $f(x)$ used to draw the curve. It came as an amazing surprise to the world of mathematics that the process of finding the slope of a tangent line and the process of finding the area under a curve were exact opposites. In order to find the area under a curve $y = f(x)$, we need to find a function $g(x)$ whose derivative is $f(x)$.

We can use this to evaluate $\int_a^b f(x)\,dx$ by the Fundamental Theorem of Calculus (see Figure 17.6).

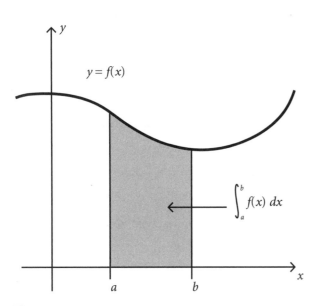

$y = f(x)$

$\int_a^b f(x)\ dx$

**Figure 17.6**

For example, the derivative of $g(x) = x^2$ is $g'(x) = 2x$. Thus, the area under $f(x) = 2x$ between $x = 3$ and $x = 5$ is $\int_3^5 2x\,dx = g(5) - g(3) = 5^2 - 3^2 = 16$. This is exactly the area of the trapezoid under the line $y = 2x$ between $x = 3$ and $x = 5$. However, here the Fundamental Theorem of Calculus saves us from having to draw out the graph of $y = 2x$.

## Example

The derivative of $g(x) = \dfrac{1}{3}x^3$ is $g'(x) = x^2$. Use this to evaluate $\int_{-1}^2 x^2\,dx$.

## Solution

By the Fundamental Theorem of Calculus, $\int_a^b f(x)\,dx = g(b) - g(a)$, where $g'(x) = f(x)$. Thus,

$$\int_{-1}^2 x^2\,dx = g(2) - g(-1)$$

$$= \frac{1}{3}(2)^3 - \frac{1}{3}(-1)^3 = \frac{8}{3} + \frac{1}{3} = 3\,.$$

Even if we *had* drawn out the graph of $y = x^2$, how would we have been able to guess that the area of the two shaded curves add up to exactly three? This is why the Fundamental Theorem of Calculus is so powerful! (See Figure 17.7.)

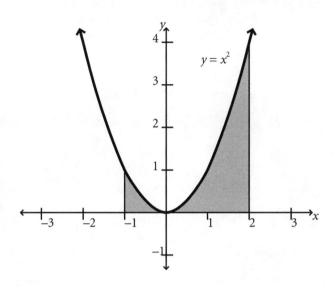

**Figure 17.7**

## Example

If $g(x) = x^4$, then $g'(x) = 4x^3$. Use this to evaluate $\int_{-1}^{1} 4x^3 \, dx$.

## Solution

$$\int_{-1}^{1} 4x^3 \, dx = g(1) - g(-1)$$

$$= 1^4 - (-1)^4 = 1 - 1 = 0$$

The answer is zero because there is exactly as much area above the $x$-axis (which counts positively) as there is below the $x$-axis (which counts negatively).

## ▶ Practice

If $g(x) = x^2 + x$ then $g'(x) = 2x + 1$. Use this to evaluate the following.

**13.** $\int_{1}^{3} (2x + 1) \, dx$

**14.** $\int_{-3}^{1} (2x + 1) \, dx$

**15.** $\int_{2}^{6} (2x + 1) \, dx$

**16.** $\int_{0}^{4} (2x + 1) \, dx$

Use $\dfrac{d}{dx}\left(\dfrac{2}{3}x^{\frac{3}{2}}\right) = \sqrt{x}$ to evaluate the following.

**17.** $\int_{0}^{1} \sqrt{x} \, dx$

**18.** $\int_{0}^{4} \sqrt{x} \, dx$

**19.** $\int_{4}^{9} \sqrt{x} \, dx$

**20.** $\int_{0}^{100} \sqrt{x} \, dx$

Use $\dfrac{d}{dx}\left(-\dfrac{1}{x}\right) = \dfrac{1}{x^2}$ to evaluate the following.

**21.** $\int_{1}^{2} \dfrac{1}{x^2} \, dx$

**22.** $\int_{1}^{5} \dfrac{1}{x^2} \, dx$

**23.** $\int_{2}^{4} \dfrac{1}{x^2} \, dx$

**24.** $\int_{-3}^{-1} \dfrac{1}{x^2} \, dx$

LESSON

# 18 ▶ Antidifferentiation

The Fundamental Theorem of Calculus shows that the area under the curve, $\int_a^b f(x)\,dx$, can be calculated with a function $g(x)$ whose derivative is $g'(x) = f(x)$:

$$\int_a^b f(x)\,dx = \big[g(x)\big]_a^b = g(b) - g(a)$$

Because of this, the symbol $\int$, without the limits of integration, is used to represent the opposite of taking the derivative. An integral like $\int_a^b f(x)\,dx$ is called a *definite integral* because it represents a definite area. An integral like $\int f(x)\,dx$ is called an *indefinite integral* because it represents another function.

Thus, $\int f(x)\,dx$ means the *antiderivative* of $f(x)$ or "the function whose derivative is $f(x)$." For example, $\int 2x\,dx$ asks "whose derivative is $2x$?" This could be $x^2$ because $\frac{d}{dx}(x^2) = 2x$. However, it could also be $x^2 + 5$ because $\frac{d}{dx}(x^2 + 5) = 2x$ as well. In fact, because the derivative of a constant is zero, $\int 2x\,dx$ could be $x^2$ plus any constant. Therefore, we write $\int 2x\,dx = x^2 + c$ where $c$ is any constant.

The brackets $\left[\dots\right]_a^b$ are just a way of keeping track of the *limits of integration a* and *b* before they are plugged into $g(x)$ and subtracted.

Again, because mathematicians are lazy, we usually simply write $\int 2x\,dx = x^2 + c$ and assume that everyone knows that the $c$ stands for "some constant." In many ways, the "plus $c$" is the trademark of the indefinite integral because every problem that begins with $\int(\dots)\,dx$ ends with $+ c$.

If we are dealing with a definite integral like $\int_3^5 2x\,dx$, then it does not matter what constant we use. For example:

$$\int_3^5 2x\,dx = \left[x^2 + c\right]_3^5$$

$$= (5^2 + c) - (3^2 + c)$$

$$= 25 + c - 9 - c = 16$$

The "plus $c$" will always cancel out in the subtraction, so we may as well simply use $c = 0$ and write:

$$\int_3^5 2x\,dx = \left[x^2\right]_3^5 = 5^2 - 3^2 = 25 - 9 = 16$$

### Example

Use $\frac{d}{dx}(x^3 + 10x^2 + 3x) = 3x^2 + 20x + 3$ to

evaluate $\int(3x^2 + 20x + 3)\,dx$ and

$\int_1^2(3x^2 + 20x + 3)\,dx$.

### Solution

Because $\frac{d}{dx}(x^3 + 10x^2 + 3x) = 3x^2 + 20x + 3$, we know that:

$$\int(3x^2 + 20x + 3)\,dx = x^3 + 10x^2 + 3x + c$$

Similarly,

$$\int_1^2(3x^2 + 20x + 3)\,dx = \left[x^3 + 10x^2 + 3x\right]_1^2$$

$$\int_1^2(3x^2 + 20x + 3)\,dx =$$

$$((2)^3 + 10\cdot(2)^2 + 3\cdot(2)) -$$

$$((1)^3 + 10\cdot(1) + 3\cdot(1))$$

$$\int_1^2(3x^2 + 20x + 3)\,dx =$$

$$8 + 40 + 6 - (1 + 10 + 3) = 54 - 14 = 40$$

The general rules for antiderivatives are fairly simple. To take the derivative of $f(x) = x^5$, we first multiply by the exponent 5, and then we subtract one from the exponent. Thus, $f'(x) = 5x^4$.

To antidifferentiate $\int 5x^4 dx$, we must do the exact opposite of this process. First, we add one to the exponent, and then we divide the result by the new

You can verify your answer by taking its derivative. If the derivative of your answer is what you were trying to integrate, then you are correct.

The derivative of $\frac{2}{3}x^{\frac{3}{2}} + c$ is $\frac{d}{dx}\left(\frac{2}{3}x^{\frac{3}{2}} + c\right) = \frac{2}{3}\cdot\frac{3}{2}x^{\frac{1}{2}} + 0 = x^{\frac{1}{2}} = \sqrt{x}$. This verifies that $\int\sqrt{x}\,dx = \frac{2}{3}x^{\frac{3}{2}} + c$.

exponent. Thus, $\int 5x^4 dx = \frac{5x^{4+1}}{4+1} + c = x^5 + c$. In general, we write:

$$\int x^n dx = \frac{x^{n+1}}{n+1} + c \text{ if } n \neq -1$$

**Example**

Evaluate $\int x^7 dx$.

**Solution**

$$\int x^7 dx = \frac{x^{7+1}}{7+1} + c = \frac{1}{8}x^8 + c$$

**Example**

Evaluate $\int \sqrt{x}\,dx$.

**Solution**

$$\int \sqrt{x}\,dx = \int x^{\frac{1}{2}}dx$$
$$= \frac{x^{\frac{1}{2}+1}}{\frac{1}{2}+1} + c = \frac{x^{\frac{3}{2}}}{\frac{3}{2}} + c = \frac{2}{3}x^{\frac{3}{2}} + c$$

**Example**

Evaluate $\int_0^2 x^3 dx$.

**Solution**

$$\int_0^2 x^3 dx = \left[\frac{1}{4}x^4\right]_0^2$$
$$\int_0^2 x^3 dx = \frac{1}{4}\cdot 2^4 - \frac{1}{4}\cdot 0^4$$
$$= \frac{1}{4}\cdot 16 - \frac{1}{4}\cdot 0 = 4 - 0 = 4$$

## ▶ Practice

Evaluate the following integrals.

1. $\int x^5 dx$

2. $\int x^{12} dx$

3. $\int u^6 du$

4. $\int_0^6 x^2 dx$

5. $\int_1^9 x\,dx$

6. $\int t^{-3} dt$

**7.** $\displaystyle\int_{-1}^{2} t^5\, dt$

**8.** $\displaystyle\int x^{\frac{5}{3}}\, dx$

**9.** $\displaystyle\int \sqrt[3]{x}\, dx$

**10.** $\displaystyle\int \sqrt[3]{u}\, du$

**11.** $\displaystyle\int 5\, dx$

**12.** $\displaystyle\int 5\, dt$

**13.** $\displaystyle\int_{-1}^{4} 8\, dx$

**14.** $\displaystyle\int_{2}^{4} \frac{1}{x^2}\, dx$

Just as with derivatives, constants can stand aside, and the terms of sums can be dealt with separately.

**Example**

Evaluate $\displaystyle\int 5x^2\, dx$.

**Solution**

$$\int 5x^2\, dx = 5\left(\frac{1}{3}x^3\right) + c$$

$$\int 5x^2\, dx = \frac{5}{3}x^3 + c$$

**Example**

Evaluate $\displaystyle\int (2t^3 - 8t + 7)\, dt$.

**Solution**

$$\int (2t^3 - 8t + 7)\, dt = 2\cdot\frac{1}{4}t^4 - 8\cdot\frac{1}{2}t^2 + 7t + c$$

$$\int (2t^3 - 8t + 7)\, dt = \frac{1}{2}t^4 - 4t^2 + 7t + c$$

**Example**

Evaluate $\displaystyle\int_{1}^{4}\left(6\sqrt{x} - \frac{8}{x^5}\right)dx$.

**Solution**

It always helps to write everything in exponential form.

$$\int_{1}^{4}\left(6\sqrt{x} - \frac{8}{x^5}\right)dx = \int_{1}^{4}\left(6x^{\frac{1}{2}} - 8x^{-5}\right)dx$$

$$\int_{1}^{4}\left(6\sqrt{x} - \frac{8}{x^5}\right)dx = \left[6\cdot\left(\frac{2}{3}x^{\frac{3}{2}}\right) - 8\left(\frac{x^{-4}}{-4}\right)\right]_{1}^{4}$$

$$\int_{1}^{4}\left(6\sqrt{x} - \frac{8}{x^5}\right)dx = \left[4x^{\frac{3}{2}} + \frac{2}{x^4}\right]_{1}^{4}$$

$$\int_{1}^{4}\left(6\sqrt{x} - \frac{8}{x^5}\right)dx =$$

$$\left(4(8) + \frac{2}{256}\right) - \left(4 + \frac{2}{1}\right) = 26 + \frac{1}{128}$$

## ▶ Practice

Evaluate the following integrals.

**15.** $\displaystyle\int 9x^4\,dx$

**16.** $\displaystyle\int 8u^2\,du$

**17.** $\displaystyle\int (x - \sqrt{x})\,dx$

**18.** $\displaystyle\int (6x^2 - 10x + 5)\,dx$

**19.** $\displaystyle\int_0^2 12x^3\,dx$

**20.** $\displaystyle\int_1^2 (3x^2 + 4)\,dx$

**21.** $\displaystyle\int_2^7 (6x - 4)\,dx$

**22.** $\displaystyle\int (3t^{11} + 9t^2 + t)\,dt$

**23.** $\displaystyle\int_0^3 (1 - t^2)\,dt$

**24.** $\displaystyle\int (8x^3 + 10x^2 - 4x + 2)\,dx$

**25.** $\displaystyle\int_0^2 (10u^4 - 4u + 1)\,du$

**26.** $\displaystyle\int_1^9 12\sqrt{x}\,dx$

**27.** $\displaystyle\int \frac{4}{x^3}\,dx$

**28.** $\displaystyle\int (3x^{\frac{10}{3}} - 8x^{\frac{4}{7}})\,dx$

The integrals of $e^x$, $\sin(x)$, and $\cos(x)$ follow directly from their derivatives:

$$\int e^x\,dx = e^x + c \text{ because } \frac{d}{dx}(e^x) = e^x$$

$$\int \cos(x)\,dx = \sin(x) + c \text{ because}$$

$$\frac{d}{dx}(\sin(x)) = \cos(x)$$

$$\int \sin(x)\,dx = -\cos(x) + c \text{ because}$$

$$\frac{d}{dx}(-\cos(x)) = \sin(x)$$

The integral of $\ln(x)$ will have to wait until Lesson 20, though we can use the fact that $\frac{d}{dx}(\ln(x)) = \frac{1}{x}$ right now. We are inclined to say that $\int \frac{1}{x}\,dx = \ln(x) + c$, but this is not entirely correct. The derivative of $\ln(-x)$ is $\frac{d}{dx}(\ln(-x)) = \frac{1}{-x} \cdot \frac{d}{dx}(-x) = \frac{1}{-x} \cdot (-1) = \frac{1}{x}$ as well. It does not matter if the $x$ inside the natural logarithm is positive or negative, so we can generalize with the absolute value $|x|$.

$$\int \frac{1}{x}\,dx = \ln|x| + c$$

Incidentally, this nicely fills a hole in an earlier formula:

$$\int x^n\,dx = \frac{x^{n+1}}{n+1} + c \text{ if } n \neq -1$$

and if $n = -1$ then

$$\int x^{-1} dx = \int \frac{1}{x} dx = \ln|x| + c$$

**Example**

Evaluate $\int (3\sin(x) + 5\cos(x)) dx$.

**Solution**

$$\int (3\sin(x) + 5\cos(x)) dx =$$

$$-3\cos(x) + 5\sin(x) + c$$

**Example**

Evaluate $\int_0^1 (3t^2 - 5e^t) dt$.

**Solution**

$$\int_0^1 (3t^2 - 5e^t) dt = [t^3 - 5e^t]_0^1$$

$$\int_0^1 (3t^2 - 5e^t) dt = (1^3 - 5e^1) - (0^3 - 5e^0)$$

$$\int_0^1 (3t^2 - 5e^t) dt = 1 - 5e + 5 = 6 - 5e$$

**Example**

Evaluate $\int \left( x^2 + x + 1 + \frac{1}{x} + \frac{1}{x^2} \right) dx$.

**Solution**

$$\int \left( x^2 + x + 1 + \frac{1}{x} + \frac{1}{x^2} \right) dx =$$

$$\int (x^2 + x^1 + x^0 + x^{-1} + x^{-2}) dx$$

$$\int \left( x^2 + x + 1 + \frac{1}{x} + \frac{1}{x^2} \right) dx =$$

$$\frac{1}{3}x^3 + \frac{1}{2}x^2 + x^1 + \ln|x| - x^{-1} + c$$

$$\int \left( x^2 + x + 1 + \frac{1}{x} + \frac{1}{x^2} \right) dx =$$

$$\frac{1}{3}x^3 + \frac{1}{2}x^2 + x + \ln|x| - \frac{1}{x} + c$$

▶ **Practice**

Evaluate the following integrals.

**29.** $\int (x^2 - 5\cos(x)) dx$

**30.** $\int (3e^x + 2x^3) dx$

**31.** $\int \frac{1}{u} du$

**32.** $\int (\theta + 2\sin(\theta)) d\theta$

**33.** $\int (\sin(x) + 2e^x) dx$

**34.** $\int_0^1 (x + e^x) dx$

**35.** $\int_1^e \frac{4}{x} dx$

**36.** $\int_0^{\frac{\pi}{4}} (8\cos(x)) dx$

# 19 ▶ Integration by Substitution

The opposite of the Chain Rule is an integration technique called *substitution*. Using the Chain Rule, for example, the derivative of $8(3x^2 + 7)^5$ is $\frac{d}{dx}(8(3x^2 + 7)^5) = 8 \cdot 5(3x^2 + 7)^4 \cdot 6x = 240x(3x^2 7)^4$. The corresponding antiderivative is thus $\int 240x(3x^2 + 7)^4 dx = 8(3x^2 + 7)^5 + c$. It is easy to recognize this after seeing the derivative worked out, but how should we know this otherwise?

The mantra of the Chain Rule is "multiply by the derivative of the inside." So the first step to undoing it is to identify what "the inside" must have been. We substitute a new variable $u$ for this and then try to rewrite the whole integral in terms of $u$.

For example, when confronted by $\int 240x(3x^2 + 7)^4 dx$, we first notice that this is not an easy integral to solve. If we multiplied out the fourth power, then it would be a polynomial that we know how to evaluate, but that would be quite difficult. Instead, we guess that "the inside" is the stuff inside the parentheses, and substitute $u = 3x^2 + 7$.

To convert the integral entirely over to $u$, we will need to replace the $dx$ into a $du$. Because $u = 3x^2 + 7$, we know that $\dfrac{du}{dx} = 6x$. It is technically wrong to cross-multiply and say $du = 6x\,dx$, but this does result in the correct answer, so we'll go with it. Thus, $dx = \dfrac{du}{6x}$. The process of substitution works as follows.

Start with the original integral.

$$\int 240x(3x^2 + 7)^4\,dx$$

Substitute $u = 3x^2 + 7$ and $dx = \dfrac{du}{6x}$.

$$\int 240x(u)^4 \frac{du}{6x}$$

Simplify.

$$\int 40u^4\,du$$

Evaluate.

$$8u^5 + c$$

Replace $u = 3x^2 + 7$.

$$8(3x^2 + 7)^5 + c$$

Thus, $\displaystyle\int 240x(3x^2 + 7)^4\,dx = 8(3x^2 + 7)^5 + c$, as we already knew.

In general, try using something inside parentheses with $u$. If every $x$ doesn't cancel out when replacing $dx$ with $du$, then try using something else as $u$. Sometimes, the entire denominator can be used as $u$. Sometimes, nothing works and a different technique must be tried.

## Example

Evaluate $\displaystyle\int x^2\sin(x^3)\,dx$.

## Solution

If we use the stuff inside the only set of parentheses, then $u = x^3$, and thus $du = 3x^2\,dx$ and $dx = \dfrac{du}{3x^2}$.

Start with the original integral.

$$\int x^2\sin(x^3)\,dx$$

Substitute $u = x^3$ and $dx = \dfrac{du}{3x^2}$.

$$\int x^2\sin(u)\frac{du}{3x^2}$$

Simplify.

$$\int \frac{1}{3}\sin(u)\,du$$

Every $x$ is gone, so we can evaluate.

$$-\frac{1}{3}\cos(u) + c$$

Replace $u = x^3$.

$$-\frac{1}{3}\cos(x^3) + c$$

Thus, $\displaystyle\int x^2\sin(x^3)\,dx = -\frac{1}{3}\cos(x^3) + c$. This can be verified by differentiating $\dfrac{d}{dx}\left(-\dfrac{1}{3}\cos(x^3) + c\right) = -\dfrac{1}{3}(-\sin(x^3)\cdot 3x^2) + 0 = x^2\sin(x^3)$.

If we had been faced with $\int \sin(x^3)\,dx$ in the last example, then substituting $u = x^3$ would have resulted in $\int \sin(u)\dfrac{du}{3x^2}$. This cannot be evaluated because it is not entirely in terms of $u$. In fact, this integral is very difficult to solve and requires the advanced technique of replacing $\sin(x^3)$ with an infinitely long polynomial called a *power series*. Many such integrals exist that are difficult to solve, and some have completely baffled every effort to solve them so far. This book will focus on the ones that can be evaluated with basic techniques.

## Example

Evaluate $\int \dfrac{3}{2x + 7}\,dx$.

## Solution

Because there are no parentheses, try using the denominator: $u = 2x + 7$. Here, $\dfrac{du}{dx} = 2$, so $du = 2\,dx$ and $dx = \dfrac{du}{2}$.

Start with the original integral.

$$\int \frac{3}{2x + 7}\,dx$$

Substitute $u = 2x + 7$ and $dx = \dfrac{du}{2}$.

$$\int \frac{3}{u}\frac{du}{2}$$

Simplify.

$$\int \frac{3}{2}\cdot\frac{1}{u}\,du$$

Every $x$ is gone, so we can evaluate.

$$\frac{3}{2}\ln|u| + c$$

Replace $u = 2x + 7$.

$$\frac{3}{2}\ln|2x + 7| + c$$

Thus, $\int \dfrac{3}{2x + 7}\,dx = \dfrac{3}{2}\ln|2x + 7| + c$.

Basically, the dream is to find a $u$ whose derivative is elsewhere in the integral, so that between the $u$ and the $du$, every $x$ goes away. This leads to some clever tricks, as will be demonstrated in the following examples.

## Example

Evaluate $\int \dfrac{\ln(x)}{x}\,dx$.

## Solution

Here, we use $u = \ln(x)$. This is not because it is in parentheses but because its derivative $\dfrac{du}{dx} = \dfrac{1}{x}$ makes up the rest of the integral. Here, $du = \dfrac{1}{x}\,dx$, so $dx = x\,du$.

Start with the original integral.

$$\int \frac{\ln(x)}{x}\,dx$$

Substitute $u = \ln(x)$ and $dx = x\,du$.

$$\int \frac{u}{x}(x\,du)$$

Simplify.

$$\int u\,du$$

Every $x$ is gone, so we can evaluate.

$$\frac{1}{2}u^2 + c$$

Replace $u = \ln(x)$.

$$\frac{1}{2}(\ln(x))^2 + c$$

Thus, $\displaystyle\int \frac{\ln(x)}{x}\,dx = \frac{1}{2}(\ln(x))^2 + c$.

## Example

Evaluate $\displaystyle\int \sin(x)\cos^3(x)\,dx$.

## Solution

Here, the trick is to use $u = \cos(x)$ so that $\dfrac{du}{dx} = -\sin(x)$ and $dx = -\dfrac{du}{\sin(x)}$.

Start with the original integral.

$$\int \sin(x)\cos^3(x)\,dx$$

Substitute $u = \cos(x)$ and $dx = -\dfrac{du}{\sin(x)}$.

$$\int \sin(x)\cdot u^3\left(-\frac{du}{\sin(x)}\right)$$

Simplify.

$$\int -u^3\,du$$

Every $x$ is gone, so we can evaluate.

$$-\frac{1}{4}u^4 + c$$

Replace $u = \cos(x)$.

$$-\frac{1}{4}\cos^4(x) + c$$

Thus, $\displaystyle\int \sin(x)\cos^3(x)\,dx = -\frac{1}{4}\cos^4(x) + c$.

To use substitution on a definite integral, it is best to evaluate the indefinite integral first.

## Example

Evaluate $\displaystyle\int_1^5 \sqrt{3x+1}\,dx$.

## Solution

First, we evaluate $\displaystyle\int \sqrt{3x+1}\,dx$ using $u = 3x+1$, $du = 3\,dx$, and $dx = \dfrac{du}{3}$.

Start with the original integral.

$$\int \sqrt{3x+1}\,dx$$

Substitute $u = 3x+1$ and $dx = \dfrac{du}{3}$.

$$\int \sqrt{u}\,\frac{du}{3}$$

Simplify.

$$\int \frac{1}{3}u^{\frac{1}{2}}\,du$$

Every $x$ is gone, so we can evaluate.

$$\frac{1}{3} \cdot \frac{2}{3} u^{\frac{3}{2}} + c$$

Replace $u = 3x + 1$.

$$\frac{2}{9}(3x + 1)^{\frac{3}{2}} + c$$

Because $\displaystyle\int \sqrt{3x + 1}\, dx = \frac{2}{9}(3x + 1)^{\frac{3}{2}} + c$, it follows that $\displaystyle\int_1^5 \sqrt{3x + 1}\, dx = \left[ \frac{2}{9}(3x + 1)^{\frac{3}{2}} \right]_1^5 =$

$$\frac{2}{9}(16)^{\frac{3}{2}} - \frac{2}{9}(4)^{\frac{3}{2}} = \frac{2}{9}(64 - 8) = \frac{112}{9}.$$

If the wrong $u$ is chosen, then either some of the variables $x$ will still remain or else the simplified integral will still be hard to solve. If this happens, go back to the beginning and try a different $u$. Don't forget that many integrals, like those of the previous lesson, don't require substitution at all. Like much of mathematics, integration often requires patience and a knack that is developed with practice.

## ▶ Practice

Evaluate the following integrals.

**1.** $\displaystyle\int x^4(x^5 + 1)^7\, dx$

**2.** $\displaystyle\int (4x + 3)^{10}\, dx$

**3.** $\displaystyle\int_0^1 x^2(x^3 - 1)^4\, dx$

**4.** $\displaystyle\int (x^3 - 9x + 4)\, dx$

**5.** $\displaystyle\int x\sqrt{x^2 - 1}\, dx$

**6.** $\displaystyle\int_1^4 3\sqrt{x}\, dx$

**7.** $\displaystyle\int_0^7 \sqrt{3x + 4}\, dx$

**8.** $\displaystyle\int \frac{9x^2 - 5}{3x^3 - 5x}\, dx$

**9.** $\displaystyle\int 2x^3\cos(x^4)\, dx$

**10.** $\displaystyle\int \frac{6x^3 - 1}{\sqrt{3x^4 - 2x + 1}}\, dx$

**11.** $\displaystyle\int (8x + 5)(4x^2 + 5x - 1)^3\, dx$

**12.** $\displaystyle\int \frac{x}{(4x^2 + 5)^3}\, dx$

**13.** $\displaystyle\int \frac{1}{4x + 10}\, dx$

**14.** $\displaystyle\int \sin(x)\cos(x)\, dx$

**15.** $\displaystyle\int \sin^2(x)\cos(x)\, dx$

**16.** $\displaystyle\int \cos(4x)\, dx$

**17.** $\displaystyle\int 4\cos(x)\, dx$

**18.** $\displaystyle\int \sin(7x - 2)\, dx$

**19.** $\displaystyle\int e^x\sin(e^x)\, dx$

**20.** $\displaystyle\int \frac{(\ln(x))^3}{x}\,dx$

**21.** $\displaystyle\int \frac{1}{x\ln(x)}\,dx$

**22.** $\displaystyle\int xe^{(x^2)}\,dx$

**23.** $\displaystyle\int \tan(x)\,dx = \int \frac{\sin(x)}{\cos(x)}\,dx$

**24.** $\displaystyle\int \frac{e^x}{1 + e^x}\,dx$

# 20 ▶ Integration by Parts

The integral of the product of two things is unfortunately not the product of the integrals. For example, the integral $\int x \cdot \cos(x)\,dx$ is not $\frac{1}{2}x^2\sin(x) + c$. We know this because the derivative of $\frac{1}{2}x^2\sin(x) + c$ is, by the Product Rule, $\frac{d}{dx}\left(\frac{1}{2}x^2\sin(x) + c\right) = 2x \cdot \sin(x) + \cos(x) \cdot \frac{1}{2}x^2$, which is not equal to $x \cdot \cos(x)$. It is unfortunate that this does not work because, if it did, evaluating integrals would be simple and would not require so many different techniques.

The integration technique that undoes the Product Rule is called *integration by parts*. The derivative of $u \cdot v$, using the Product Rule, can be expressed as $du \cdot v + dv \cdot u$ or $u\,dv + v\,du$. The corresponding integral is:

$$\int (u\,dv + v\,du) = uv$$

This can be broken up into $\int u\,dv + \int v\,du = uv$ and written as:

Integration by Parts Formula: $\int u\,dv = uv - \int v\,du$

A good mnemonic for guessing what to use as $u$ is "LIPET." That is, let $u$ be a **Logarithm** if there is one. If not, let $u$ be the **Inverse** of a trigonometric function (not covered in this book). If there isn't either of these, then let $u$ be a **Polynomial**, and if there is none, let it be an **Exponential** function. Only as the very last resort, should you let $u$ be a **Trigonometric** function.

This can often be used to transform a difficult integral into one that is solvable. For example, take $\int x \cdot \cos(x)\,dx$. This looks just like $\int u\,dv$ if $u = x$ and $dv = \cos(x)\,dx$. In order to use the formula, we will need to get $du$ by differentiating $u$. Because $u = x$, we know that $\frac{du}{dx} = 1$, so $du = dx$. We will also need to get $v$ from $dv$ by integrating. And because $dv = \cos(x)\,dx$, it must be that $v = \sin(x)$. Thus:

$$\int x \cdot \cos(x)\,dx = \int u\,dv$$

$$\int x \cdot \cos(x)\,dx = uv - \int v\,du$$

$$\int x \cdot \cos(x)\,dx = x\sin(x) - \int \sin(x)\,dx$$

$$\int x \cdot \cos(x)\,dx = x\sin(x) + \cos(x) + c$$

This is the correct answer, as can be verified by taking the derivative $\frac{d}{dx}(x\sin(x) + \cos(x) + c) = 1 \cdot \sin(x) + \cos(x) \cdot x - \sin(x) + 0 = x \cdot \cos(x)$.

## Example

Evaluate $\int xe^x\,dx$.

## Solution

This cannot be solved by basic integration or by substitution, so we try integration by parts. No logarithms or inverse trigonometric functions are found here, but there is the polynomial $x$, so we try $u = x$. The $dv$ must then be everything else after the integral sign, so $dv = e^x\,dx$. After differentiating $u$ and integrating $dv$, we get:

$$u = x$$

$$du = dx$$

And:

$$dv = e^x\,dx$$

$$v = e^x$$

Thus, using the integration by parts formula $\int u\,dv = uv - \int v\,du$, we evaluate as follows:

$$\int xe^x\,dx = \int u\,dv$$

$$\int xe^x\,dx = uv - \int v\,du$$

$$\int xe^x\,dx = xe^x - \int e^x\,dx$$

$$\int xe^x\,dx = xe^x - e^x + c$$

## Example

Evaluate $\int x^3 \ln(x)\, dx$.

## Solution

Here, we have a logarithm, so we set $u = \ln(x)$ and $dv = x^3\, dx$. Thus:

$$u = \ln(x)$$

$$du = \frac{1}{x}\, dx$$

And:

$$dv = x^3\, dx$$

$$v = \frac{1}{4}x^4$$

And then we evaluate.

$$\int x^3 \ln(x)\, dx = \int u\, dv$$

$$\int x^3 \ln(x)\, dx = uv - \int v\, du$$

$$\int x^3 \ln(x)\, dx = \ln(x) \cdot \frac{1}{4}x^4 - \int \frac{1}{4}x^4 \cdot \frac{1}{x}\, dx$$

$$\int x^3 \ln(x)\, dx = \frac{1}{4}x^4 \ln(x) - \int \frac{1}{4}x^3\, dx$$

$$\int x^3 \ln(x)\, dx = \frac{1}{4}x^4 \ln(x) - \frac{1}{16}x^4 + c$$

This can even solve the following problem that was mentioned in Lesson 18.

## Example

Evaluate $\int \ln(x)\, dx$.

## Solution

Because there seems to be only one part to this integral, one wouldn't think to try integration by parts first. However, because nothing else will work, we can try $u = \ln(x)$. The only thing left for the $dv$ is $dx$, so we use $dv = dx$, which leads to $v = x$.

$$u = \ln(x)$$

$$du = \frac{1}{x}\, dx$$

And:

$$dv = dx$$

$$v = x$$

And now evaluate as follows.

$$\int \ln(x)\, dx = \int u\, dv$$

$$\int \ln(x)\, dx = uv - \int v\, du$$

$$\int \ln(x)\, dx = \ln(x) \cdot x - \int x \cdot \frac{1}{x}\, dx$$

$$\int \ln(x)\, dx = x\ln(x) - \int 1\, dx$$

$$\int \ln(x)\, dx = x\ln(x) - x + c$$

Sometimes, integration by parts needs to be done more than once to solve a problem.

## Example

Evaluate $\int x^2 \cos(x)\, dx$.

## Solution

Here, $u = x^2$, so $du = 2x\,dx$, and $dv = \cos(x)\,dx$, so $v = \sin(x)$.

$$\int x^2\cos(x)\,dx = \int u\,dv$$

$$\int x^2\cos(x)\,dx = uv - \int v\,du$$

$$\int x^2\cos(x)\,dx = x^2\sin(x) - \int \sin(x) \cdot 2x\,dx$$

$$\int x^2\cos(x)\,dx = x^2\sin(x) - \int 2x\sin(x)\,dx$$

In order to solve this $\int 2x\sin(x)\,dx$, we have to use integration by parts a second time, but this time, with $u = 2x$ and $dv = \sin(x)\,dx$.

$$u = 2x$$

$$du = 2\,dx$$

And:

$$dv = \sin(x)\,dx$$

$$v = -\cos(x)$$

Now we evaluate as follows.

$$\int x^2\cos(x)\,dx = x^2\sin(x) - \int 2x\sin(x)\,dx$$

$$\int x^2\cos(x)\,dx = x^2\sin(x) - \int u\,dv$$

$$\int x^2\cos(x)\,dx = x^2\sin(x) - \left(uv - \int v\,du\right)$$

$$\int x^2\cos(x)\,dx =$$

$$x^2\sin(x) - \left(2x \cdot (-\cos(x)) - \int (-\cos(x)) \cdot 2\,dx\right)$$

$$\int x^2\cos(x)\,dx =$$

$$x^2\sin(x) + 2x\cos(x) + \int -2\cos(x)\,dx$$

$$\int x^2\cos(x)\,dx =$$

$$x^2\sin(x) + 2x\cos(x) - 2\sin(x) + c$$

The final example utilizes a clever trick that few people have ever figured out on their own. Instead, they have seen it done and learned to copy it. Opportunities to use this trick are few, but it is interesting enough to see at least once.

## Example

Evaluate $\int e^x\sin(x)\,dx$.

## Solution

The first letter of LIPET that we reach is E because there are neither logarithms nor polynomials, so let $u = e^x$ and $dv = \sin(x)\,dx$.

$$u = e^x$$

$$du = e^x dx$$

And:

$$dv = \sin(x)\,dx$$

$$v = -\cos(x)$$

And now evaluate as follows.

$$\int e^x \sin(x)\,dx = \int u\,dv$$

$$\int e^x \sin(x)\,dx = uv - \int v\,du$$

$$\int e^x \sin(x)\,dx =$$

$$e^x(-\cos(x)) - \int(-\cos(x)) \cdot e^x\,dx$$

$$\int e^x \sin(x)\,dx = -e^x\cos(x) + \int e^x\cos(x)\,dx$$

To evaluate $\int e^x\cos(x)\,dx$, we use integration by parts again:

$$u = e^x$$

$$du = e^x dx$$

And:

$$dv = \cos(x)\,dx$$

$$v = \sin(x)$$

And then the evaluation:

$$\int e^x\sin(x)\,dx = -e^x\cos(x) + \int e^x\cos(x)\,dx$$

$$\int e^x\sin(x)\,dx = -e^x\cos(x) + \int u\,dv$$

$$\int e^x\sin(x)\,dx = -e^x\cos(x) + uv - \int v\,du$$

$$\int e^x\sin(x)\,dx =$$

$$-e^x\cos(x) + e^x\sin(x) - \int \sin(x) \cdot e^x\,dx$$

$$\int e^x\sin(x)\,dx =$$

$$-e^x\cos(x) + e^x\sin(x) - \int e^x\sin(x)\,dx$$

Here is the moment of despair: To evaluate $\int e^x\sin(x)\,dx$, we need to be able to evaluate $\int e^x\sin(x)\,dx$! And yet, the trick here is to bring both integrals to one side of the equation:

$$\int e^x\sin(x)\,dx + \int e^x\sin(x)\,dx =$$

$$-e^x\cos(x) + e^x\sin(x)$$

$$2\int e^x\sin(x)\,dx = -e^x\cos(x) + e^x\sin(x)$$

$$\int e^x\sin(x)\,dx = \frac{1}{2}(-e^x\cos(x) + e^x\sin(x)) + c$$

## ▶ Practice

Evaluate the following integrals using integration by parts, substitution, or basic integration.

**1.** $\int x^5 \ln(x)\, dx$

**2.** $\int x\sin(x)\, dx$

**3.** $\int x\sin(x^2)\, dx$

**4.** $\int (x + 3)\cos(x)\, dx$

**5.** $\int \dfrac{\ln(x)}{x}\, dx$

**6.** $\int x^2\sin(x)\, dx$

**7.** $\int (x^2 + \sin(x))\, dx$

**8.** $\int x^2 e^{x^3 + 1}\, dx$

**9.** $\int x^2 e^x\, dx$

**10.** $\int (x^3 + 3x - 1)\ln(x)\, dx$

**11.** $\int (x + \ln(x))\, dx$

**12.** $\int \sqrt{x - 1}\, dx$

**13.** $\int x\sqrt{x - 1}\, dx$

**14.** $\int x e^{-x}\, dx$

**15.** $\int \sqrt{x}\ln(x)\, dx$

**16.** $\int \dfrac{\ln(x)}{x^3}\, dx$

**17.** $\int \dfrac{1}{x}\, dx$

**18.** $\int (x^2 - 1)\cos(x)\, dx$

**19.** $\int \dfrac{e^{\frac{1}{x}}}{x^2}\, dx$

**20.** $\int \sin(x)\sqrt{\cos(x)}\, dx$

**21.** $\int \sin(x) \cdot \ln(\cos(x))\, dx$

**22.** $\int e^x\cos(x)\, dx$

# Posttest

If you have completed all 20 lessons in this book, you are ready to take the posttest to measure your progress. The posttest has 50 multiple-choice questions covering the topics you studied in this book. Although the format of the posttest is similar to that of the pretest, the questions are different.

Take as much time as you need to complete the posttest. When you are finished, check your answers with the answer key that follow the posttest. Along with each answer is a number that tells you which lesson of this book teaches you about the calculus skills needed for that question. Once you know your score on the posttest, compare the results with the pretest. If you scored better on the posttest than you did on the pretest, congratulations! You have profited from your hard work. At this point, you should look at the questions you missed, if any. Do you know why you missed the question, or do you need to go back to the lesson and review the concept?

## ANSWER SHEET

1. (a) (b) (c) (d)
2. (a) (b) (c) (d)
3. (a) (b) (c) (d)
4. (a) (b) (c) (d)
5. (a) (b) (c) (d)
6. (a) (b) (c) (d)
7. (a) (b) (c) (d)
8. (a) (b) (c) (d)
9. (a) (b) (c) (d)
10. (a) (b) (c) (d)
11. (a) (b) (c) (d)
12. (a) (b) (c) (d)
13. (a) (b) (c) (d)
14. (a) (b) (c) (d)
15. (a) (b) (c) (d)
16. (a) (b) (c) (d)
17. (a) (b) (c) (d)

18. (a) (b) (c) (d)
19. (a) (b) (c) (d)
20. (a) (b) (c) (d)
21. (a) (b) (c) (d)
22. (a) (b) (c) (d)
23. (a) (b) (c) (d)
24. (a) (b) (c) (d)
25. (a) (b) (c) (d)
26. (a) (b) (c) (d)
27. (a) (b) (c) (d)
28. (a) (b) (c) (d)
29. (a) (b) (c) (d)
30. (a) (b) (c) (d)
31. (a) (b) (c) (d)
32. (a) (b) (c) (d)
33. (a) (b) (c) (d)
34. (a) (b) (c) (d)

35. (a) (b) (c) (d)
36. (a) (b) (c) (d)
37. (a) (b) (c) (d)
38. (a) (b) (c) (d)
39. (a) (b) (c) (d)
40. (a) (b) (c) (d)
41. (a) (b) (c) (d)
42. (a) (b) (c) (d)
43. (a) (b) (c) (d)
44. (a) (b) (c) (d)
45. (a) (b) (c) (d)
46. (a) (b) (c) (d)
47 (a) (b) (c) (d)
48. (a) (b) (c) (d)
49. (a) (b) (c) (d)
50. (a) (b) (c) (d)

If your score on the posttest doesn't show much improvement, take a second look at the questions you missed. Did you miss a question because of an error you made? If you can figure out why you missed the problem, then you understand the concept and simply need to concentrate more on accuracy when taking a test. If you missed a question because you did not know how to work the problem, go back to the lesson and spend more time working that type of problem. Take the time to understand basic calculus thoroughly. You need a solid foundation in basic calculus if you plan to use this information or progress to a higher level. Whatever your score on this posttest, keep this book for review and future reference.

## ▶ Posttest

**1.** Evaluate $f(-2)$ when $f(x) = x^3 - 2x$.

   **a.** $-12$

   **b.** $-10$

   **c.** $-4$

   **d.** $4$

**2.** Simplify $f(2x + 1)$ when $f(x) = x^2 + x$.

   **a.** $4x^2 + 6x + 2$

   **b.** $4x^2 + 2x + 2$

   **c.** $2x^2 + 3x$

   **d.** $2x^3 + 3x^2 + x$

**3.** Evaluate $g \circ h(x)$ when $g(x) = x^2 + 5x + 1$ and $h(x) = \dfrac{1}{x}$.

   **a.** $\dfrac{6}{x}$

   **b.** $x + 5 + \dfrac{1}{x}$

   **c.** $\dfrac{1}{x^2 + 5x + 1}$

   **d.** $\dfrac{1}{x^2} + \dfrac{5}{x} + 1$

**4.** What is the domain of $f(x) = \dfrac{\sqrt{x + 1}}{x}$?

   **a.** $x \geq -1$

   **b.** $x \neq 0$

   **c.** $x \neq -1, x \neq 0$

   **d.** $x \geq -1, x \neq 0$

Use the following graph for problems 5 and 6.

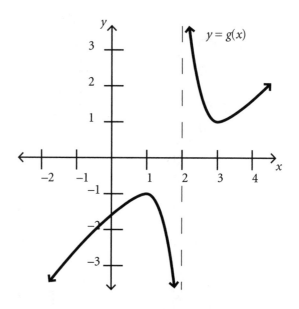

**5.** Where does $g(x)$ have a local maximum?

   **a.** $x = 0$

   **b.** $x = 1$

   **c.** $x = 2$

   **d.** $x = 3$

**6.** Where is $g(x)$ decreasing?

   **a.** $(1,2)$ and $(2,3)$

   **b.** $(-\infty,2)$

   **c.** $(-\infty,0)$

   **d.** $(0,\infty)$

**7.** What is the slope of the line through $(2,-4)$ and $(1,7)$?

   **a.** $\dfrac{1}{11}$

   **b.** $-11$

   **c.** $11$

   **d.** $3$

**8.** Simplify $4^3$.

   **a.** $7$

   **b.** $12$

   **c.** $16$

   **d.** $64$

**9.** Simplify $16^{-\frac{1}{2}}$.

   **a.** $-8$

   **b.** $4$

   **c.** $\dfrac{1}{4}$

   **d.** $\dfrac{1}{16}$

**10.** What is $x$ if $4^x = 10$?

   **a.** $\dfrac{\ln(10)}{\ln(4)}$

   **b.** $2.5$

   **c.** $10^4$

   **d.** $\ln\left(\dfrac{5}{2}\right)$

**11.** What is $\cos\left(\dfrac{\pi}{4}\right)$?

   **a.** $1$

   **b.** $\dfrac{1}{2}$

   **c.** $\dfrac{\sqrt{2}}{2}$

   **d.** $\dfrac{\sqrt{3}}{2}$

**12.** What is $\sin\left(\dfrac{4\pi}{3}\right)$?

   **a.** $-\dfrac{\sqrt{3}}{2}$

   **b.** $\dfrac{\sqrt{3}}{2}$

   **c.** $-\dfrac{\sqrt{2}}{2}$

   **d.** $\dfrac{\sqrt{2}}{2}$

**13.** Evaluate $\lim\limits_{x\to 2}\dfrac{x^2 - 5x + 6}{x^2 - 2x - 3}$.

  **a.** 0

  **b.** 1

  **c.** $\dfrac{2}{3}$

  **d.** undefined

**14.** Simplify $\lim\limits_{x\to 3}\dfrac{x^2 - 5x + 6}{x^2 - 2x - 3}$.

  **a.** $\dfrac{1}{3}$

  **b.** $\dfrac{1}{4}$

  **c.** 1

  **d.** undefined

**15.** Evaluate $\lim\limits_{x\to 1+}\dfrac{x + 4}{x^2 - 1}$.

  **a.** 5

  **b.** $-\infty$

  **c.** $\infty$

  **d.** undefined

**16.** What is the slope of the tangent line to $y = x^2$ at $x = 3$?

  **a.** 2

  **b.** 6

  **c.** 9

  **d.** $2x$

**17.** What is the slope of the tangent line to $y = 4x - 7$ at $x = 2$?

  **a.** $-7$

  **b.** $-3$

  **c.** 1

  **d.** 4

**18.** What is the derivative of $g(x) = 8x^4 - 10x^3 + 3x - 1$?

  **a.** 0

  **b.** $8x^4 - 10x^3 + 3x$

  **c.** $32x^3 - 10x^2 + 3x - x$

  **d.** $32x^3 - 30x^2 + 3$

**19.** Suppose that after $t$ seconds, a falling rock is $s(t) = -16t^2 + 5t + 200$ feet off the ground. How fast is the rock traveling after 2 seconds?

  **a.** 10 feet per second

  **b.** 59 feet per second

  **c.** 64 feet per second

  **d.** 156 feet per second

**20.** Differentiate $y = \sqrt{x} + 4\sin(x)$.

  **a.** $\dfrac{1}{2\sqrt{x}} + 4\cos(x)$

  **b.** $\sqrt{x} + 4\cos(x)$

  **c.** $\dfrac{\sqrt{x}}{2} + 4\sin(x)$

  **d.** $\dfrac{1}{2\sqrt{x}} - 4\cos(x)$

**21.** What is the derivative of $f(x) = 5e^x - 2\ln(x)$?

    **a.** $5e^x - \dfrac{2}{x}$

    **b.** $5xe^{x-1} - \dfrac{2}{x}$

    **c.** $5xe^{x-1} - 2x$

    **d.** $5e^x - 2x$

**22.** Differentiate $y = xe^x$.

    **a.** $e^x$

    **b.** $xe^x$

    **c.** $(x + 1)e^x$

    **d.** $xe^{x-1}$

**23.** Differentiate $g(x) = \dfrac{\cos(x)}{x^2 + 5x}$.

    **a.** $\dfrac{\sin(x)}{2x + 5}$

    **b.** $\dfrac{-\sin(x)}{2x + 5}$

    **c.** $\dfrac{(2x + 5)\cos(x) + (x^2 + 5)\sin(x)}{(x^2 + 5)^2}$

    **d.** $\dfrac{-(x^2 + 5x)\sin(x) - (2x + 5)\cos(x)}{(x^2 + 5x)^2}$

**24.** What is the derivative of $f(x) = \sec(x)$?

    **a.** $\sec(x)$

    **b.** $\tan^2(x)$

    **c.** $\sec(x)\tan(x)$

    **d.** $1 - \sec(x)$

**25.** What is the slope of the line that is tangent to $y = (x^2 - 2)^3$ at $x = 2$?

    **a.** 8

    **b.** 12

    **c.** 24

    **d.** 48

**26.** Differentiate $x\sin(x^2)$.

    **a.** $x\cos(x^2) + \sin(x^2)$

    **b.** $2x^2\cos(x^2) + \sin(x^2)$

    **c.** $2x^2\cos(x^2)$

    **d.** $2x\sin(x^2)$

**27.** Find $\dfrac{dy}{dx}$ when $\tan(y) + y = \ln(x) - 1$.

    **a.** $\dfrac{1}{x} - \sec^2(x)$

    **b.** $\dfrac{1}{x(1 + \sec^2(x))}$

    **c.** $\dfrac{x}{1 + \sec^2(x)}$

    **d.** $\dfrac{1}{x} - \sec^2(x) - 1$

**28.** Find $\dfrac{dy}{dx}$ when $x^2 y = xy^2$.

   **a.** $\dfrac{x}{y}$

   **b.** $\dfrac{y - 2x}{x - 2y}$

   **c.** $\dfrac{y^2 - 2xy}{x^2 - 2xy}$

   **d.** $\dfrac{y^2 + 2xy}{x^2 + 2xy}$

**29.** What is the slope of the curve $y^3 - y = 3x + 3$ at $(1,2)$?

   **a.** $\dfrac{3}{11}$

   **b.** $3$

   **c.** $\dfrac{2\sqrt{3}}{3}$

   **d.** $\dfrac{5\sqrt{3}}{3}$

**30.** The volume of a sphere is $V = \frac{4}{3}\pi r^3$. If the radius increases by 3 meters per second, how fast does the volume change when $r = 10$ meters?

   **a.** $400\pi\dfrac{\text{m}^3}{\text{sec}}$

   **b.** $\dfrac{4{,}000\pi}{3}\dfrac{\text{m}^3}{\text{sec}}$

   **c.** $4{,}000\pi\dfrac{\text{m}^3}{\text{sec}}$

   **d.** $1{,}200\pi\dfrac{\text{m}^3}{\text{sec}}$

**31.** If a 10-foot ladder slides down a wall at 2 feet per minute (see the figure that follows), how fast does the bottom slide when the top is 6 feet up?

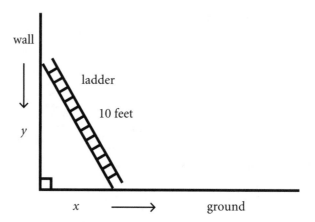

   **a.** $\dfrac{1}{2}$ foot per minute

   **b.** $\dfrac{3}{2}$ foot per minute

   **c.** 2 feet per minute

   **d.** 12 feet per minute

**32.** What is $\lim\limits_{x \to \infty} \dfrac{3x^2 + 7x - 2}{x^2 + 5x - 1}$?

   **a.** $3$

   **b.** $\dfrac{7}{5}$

   **c.** $\infty$

   **d.** undefined

**33.** Where does $y = e^x$ have a horizontal asymptote?

   **a.** $y = 0$

   **b.** $y = 1$

   **c.** $y = e$

   **d.** no asymptote

**34.** Evaluate $\displaystyle\lim_{x\to\infty} \frac{x^5 + 3x^3}{e^x - 1}$ .

  **a.** 5
  **b.** 0
  **c.** $\infty$
  **d.** undefined

**35.** On what intervals is $f(x) = x^3 + 6x^2 - 15x + 2$ decreasing?

  **a.** $(4,5)$
  **b.** $(-5,1)$
  **c.** $(2,6)$ and $(15,\infty)$
  **d.** $(-\infty,-5)$ and $(1,\infty)$

**36.** Which of the following is the graph of $y = 3x - x^3$ ?

**a.**

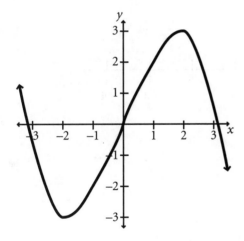

**b.**

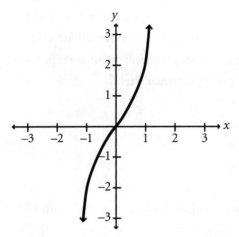

**c.**

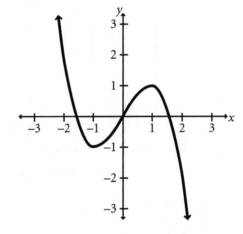

**d.**

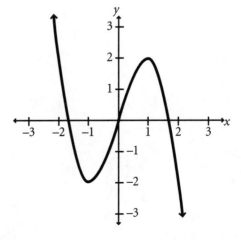

**37.** If up to 30 apple trees are planted on an acre, each will produce 400 apples a year. For every tree over 30 on the acre, each tree will produce 10 apples less each year. How many trees per acre will maximize the annual yield?

a. 5 trees

b. 32 trees

c. 35 trees

d. 40 trees

**38.** An enclosure will be built, as depicted, with 100 feet of fencing. What dimensions will maximize

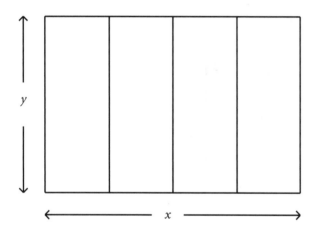

the area?

a. $x = 20, y = 12$

b. $x = 25, y = 10$

c. $x = 30, y = 8$

d. $x = 35, y = 6$

**39.** If $\int_{1}^{7} f(x)\,dx = 2$ and $\int_{7}^{10} f(x)\,dx = 8$, then what is $\int_{1}^{10} f(x)\,dx$?

a. 6

b. 9

c. 10

d. 16

**40.** What is $\int_{0}^{4} f(x)\,dx$?

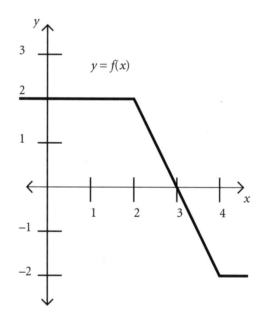

a. 0

b. 2

c. 4

d. 6

**41.** If $g(x) = \displaystyle\int_0^x t^3\, dt$, then what is $g'(x)$?

   **a.** $x^3$

   **b.** $\dfrac{1}{4}t^4 + c$

   **c.** $\dfrac{1}{4}x^4 + c$

   **d.** $3x^2$

**42.** Evaluate $\displaystyle\int_1^3 (6x^2 - 4x)\, dx$.

   **a.** 16
   **b.** 32
   **c.** 36
   **d.** 40

**43.** Evaluate $\displaystyle\int \left(x^5 - \dfrac{1}{x^2}\right) dx$.

   **a.** $\dfrac{1}{6}x^6 + \dfrac{3}{x^3} + c$

   **b.** $\dfrac{1}{6}x^6 + \dfrac{1}{x} + c$

   **c.** $5x^4 + \dfrac{2}{x^3}$

   **d.** $5x^4 + \dfrac{2}{x^3} + c$

**44.** Evaluate $\displaystyle\int \cos(x)\, dx$

   **a.** $-\sin(x) + c$

   **b.** $\sin(x) + c$

   **c.** $-\cos(x) + c$

   **d.** $\cos(x) + c$

**45.** Integrate $\displaystyle\int (3e^x - \sin(2x))\, dx$.

   **a.** $3e^x + \cos(2x) + c$

   **b.** $3e^x - \cos(2x) + c$

   **c.** $3e^x + \frac{1}{2}\cos(2x) + c$

   **d.** $\dfrac{3e^{x+1}}{x+1} - \sin(x^2) + c$

**46.** Evaluate $\displaystyle\int \dfrac{\ln(x)}{x}\, dx$.

   **a.** $\dfrac{1}{x} + c$

   **b.** $\dfrac{1}{2}x^2 + c$

   **c.** $(\ln(x))^2 + c$

   **d.** $\dfrac{1}{2}(\ln(x))^2 + c$

**47.** Evaluate $\int xe^{(x^2)}\,dx$.

   **a.** $e^{(x^2)} + c$

   **b.** $\frac{1}{2}e^{(x^2)} + c$

   **c.** $\frac{1}{2}xe^{(x^2)} + c$

   **d.** $\frac{1}{2}x^2e^{(x^2)} + c$

**48.** Evaluate $\int_0^6 \sqrt{4x + 1}\,dx$.

   **a.** $\frac{62}{3}$

   **b.** $\frac{248}{3}$

   **c.** $\frac{125}{6}$

   **d.** 124

**49.** Integrate $\int \ln(x)\,dx$.

   **a.** $\ln(1) + c$

   **b.** $\frac{1}{x} + c$

   **c.** $\frac{1}{2}(\ln(x))^2 + c$

   **d.** $x\ln(x) - x + c$

**50.** Evaluate $\int xe^x\,dx$.

   **a.** $\frac{1}{2}xe^x + c$

   **b.** $\frac{1}{2}x^2e^x + c$

   **c.** $xe^x - e^x + c$

   **d.** $xe^x + e^x + c$

# ▶ Answers

1. c. Lesson 1
2. a. Lesson 1
3. d. Lesson 1
4. d. Lesson 1
5. b. Lesson 2
6. a. Lesson 2
7. b. Lesson 2
8. d. Lesson 3
9. c. Lesson 3
10. a. Lesson 3
11. c. Lesson 4
12. a. Lesson 4
13. a. Lesson 5
14. b. Lesson 5
15. c. Lesson 5
16. b. Lessons 6, 7
17. d. Lessons 6, 7
18. d. Lesson 7
19. b. Lesson 8
20. a. Lesson 8
21. a. Lesson 8
22. c. Lesson 9
23. d. Lesson 9
24. c. Lesson 9

25. d. Lesson 10
26. b. Lessons 9, 10
27. b. Lesson 11
28. c. Lesson 11
29. a. Lesson 11
30. d. Lesson 12
31. b. Lesson 12
32. a. Lesson 13
33. a. Lesson 13
34. b. Lesson 13
35. b. Lesson 14
36. d. Lesson 14
37. c. Lesson 15
38. b. Lesson 15
39. c. Lesson 16
40. c. Lesson 16
41. a. Lesson 17
42. c. Lesson 18
43. b. Lesson 18
44. b. Lesson 18
45. c. Lessons 18, 19
46. d. Lesson 19
47. b. Lesson 19
48. a. Lesson 19
49. d. Lesson 20
50. c. Lesson 20

 Answer Key

## ► Lesson 1

**1.** $f(5) = 9$

**2.** $g(-3) = -20$

**3.** $h\left(\dfrac{1}{2}\right) = 1$

**4.** $f(7) = 2$. Because there is no $x$ in the description of $f$, the 7 never gets used. This is called a *constant* function because it is constantly equal to 2.

**5.** $k(4) = 21$

**6.** $h(64) = 4$

**7.** The rock is $s(3) = 16$ feet high after 3 seconds.

**8.** The profit on 100 cookies is $P(100) = \$39$.

**9.** $f(y) = y^2 + 3y - 1$

**10.** $f(y + 1) = y^2 + 5y + 3$

**11.** $f(x + a) = x^2 + 2xa + a^2 + 3x + 3a - 1$

**12.** $g(x^2 + \sqrt{x}) = \dfrac{8}{x^2 + \sqrt{x}} - 6(x^2 + \sqrt{x})$

**13.** $g(2x) - g(x) =$
$\dfrac{8}{2x} - 6(2x) - \left(\dfrac{8}{x} - 6x\right) = -\dfrac{4}{x} - 6x$

**14.** $f(x + a) - f(x) = (x + a)^2 +$
$4(x + a) - 5 - (x^2 +$
$4x - 5) = 2xa + a^2 + 4a$

**15.** $\dfrac{h(x + a) - h(x)}{a} =$
$\dfrac{3(x + a) + 2 - (3x + 2)}{a} = 3$

**16.** $\dfrac{g(x + a) - g(x)}{a} =$
$\dfrac{(x + a)^2 - 2(x + a) + 1 - (x^2 - 2x + 1)}{a}$
$= 2x + a - 2$

**17.** $f \circ g(x) = \dfrac{1}{x^3 - 2x^2 + 1}$

**18.** $g \circ f(x) = \dfrac{1}{x^3} - \dfrac{2}{x^2} + 1$

**19.** $f \circ h(t) = \dfrac{1}{t - \sqrt{t}}$

**20.** $f \circ f(x) = \dfrac{1}{\frac{1}{x}} = x$

**21.** $h \circ h(x) = x - \sqrt{x} - \sqrt{x - \sqrt{x}}$

**22.** $g \circ h(9) = g(9 - \sqrt{9}) = g(6) = 145$

**23.** $h \circ f \circ g(x) = h(f(g(x))) =$

$$\dfrac{1}{x^3 - 2x^2 + 1} - \sqrt{\dfrac{1}{x^3 - 2x^2 + 1}}$$

**24.** $f \circ h \circ f(2x) = f\left(h\left(\dfrac{1}{2x}\right)\right) = \dfrac{1}{\frac{1}{2x} - \sqrt{\frac{1}{2x}}}$

**25.** $x \neq -3$, $x \neq 5$

**26.** $x \geq -1$

**27.** $t > -5$ because $t \neq -5$

**28.** The domain consists of all real numbers.

**29.** $a \neq 0$

**30.** The domain consists of all real numbers.

**31.** $x \leq 2$, $x \neq -8$

**32.** $u > -\dfrac{4}{3}$ is enough, because it already rules out

$u \neq -3$ and $u \neq -\dfrac{4}{3}$.

▶ **Lesson 2**

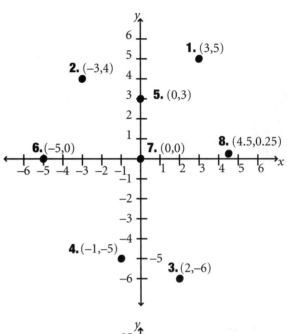

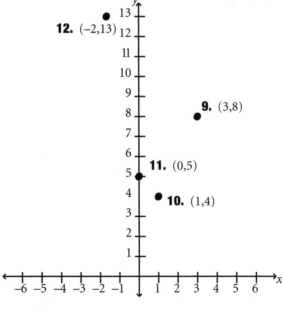

**13.** The domain of $f$ is $x \neq 0$. There is a discontinuity at $x = 0$. The graph of $f$ is decreasing on $(-\infty,0)$ and on $(0,\infty)$. The graph of $f$ is concave down on $(-\infty,0)$ and concave up on $(0,\infty)$. There are no points of inflection, no local maxima, and no local minima. There is a vertical asymptote at $x = 0$ and a horizontal asymptote at $y = 0$.

**14.** The domain of *g* consists of all real numbers. There are no discontinuities. The function increases on $(-\infty,-3)$ and on $(0,3)$, and it decreases on $(-3,0)$ and on $(3,\infty)$. There are local maxima at $(-3,4)$ and $(3,4)$, and there is a local minimum at $(2,0)$. The graph is concave up on $(-1,1)$ and concave down on $(-\infty,-1)$ and on $(1,\infty)$. There are points of inflection at $(\infty1,3)$ and $(1,3)$. There are no asymptotes.

**15.** The domain of *h* is $x \neq 1$. There is a discontinuity at $x = 1$. The function increases on $(-1,1)$ and on $(1,\infty)$, and decreases on $(-\infty,-1)$. The function is concave up on $(-\infty,1)$ and on $(1,\infty)$. There is a local minimum at $(1,-2)$. There are no asymptotes, nor any points of inflection.

**16.** The domain is $x \neq -2, 2$ with discontinuities at $x = -2$ and $x = 2$. The function increases on $(0,2)$ and $(2,\infty)$, and decreases on $(-\infty,-2)$ and on $(-2,0)$. The point $(0,2)$ is a local minimum. The graph is concave up on $(-2,2)$ and concave down on $(-\infty,-2)$ and $(2,\infty)$. There are no points of inflection. There are vertical asymptotes at $x = -2$ and $x = 2$, and a horizontal asymptote at $y = 0$.

**17.** The domain consists of all real numbers, though there is a discontinuity at $x = -1$. The function increases on $(-\infty,-1)$ and on $(-1,2)$, and it decreases on $(2,\infty)$. There are local maxima at $(-1,3)$ and $(2,3)$. The graph is concave up on $(-1,0)$ and concave down on $(0,\infty)$, so there is a point of inflection at $(0,2)$. Because the line is straight before $x = -1$, it does not curve upward or downward, and thus has no concavity. There are no asymptotes.

**18.** The domain is the whole real line, with no discontinuities. The graph increases on $(-\infty,\infty)$, is concave up on $(-\infty,0)$, and is concave down on $(0,\infty)$. There is a point of inflection at $(0,0)$. There are horizontal asymptotes at $y = -2$ and $y = 2$.

**19.** The domain is $(0,\infty)$ with no discontinuities. The graph increases on $(0,2)$, has a local maximum at $(2,5)$, and decreases on $(2,\infty)$. The graph is concave down on $(0,3)$ and concave up on $(3,\infty)$ with a point of inflection at $(3,3)$. There is a vertical asymptote at $x = 0$ and a horizontal asymptote at $y = 1$.

**20.** The domain is $x \neq 5$ with discontinuities at $x = 2$ and $x = 5$. The function increases on $(-\infty,1)$, $(4,5)$, and on $(5,\infty)$. The function decreases on $(1,2)$ and on $(2,4)$. There is a local maximum at $(1,2)$ and at $(2,3)$. The point $(4,2)$ is a local minimum. The graph is concave up on $(-\infty,1)$, $(1,2)$, $(2,5)$, and on $(5,\infty)$. There is a horizontal asymptote at $y = 0$.

**21.** 3

**22.** $\dfrac{1}{2}$

**23.** 0

**24.** $-\dfrac{9}{4}$

**25.** $\dfrac{w-7}{3} = \dfrac{7-w}{-3}$

**26.** $\dfrac{y-10}{x-4} = \dfrac{10-y}{4-x}$

**27.**

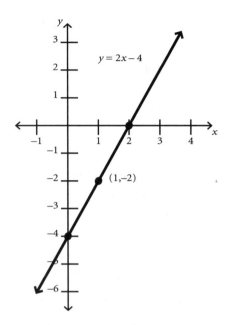

$y = 2x - 4$

$(1, -2)$

**29.**

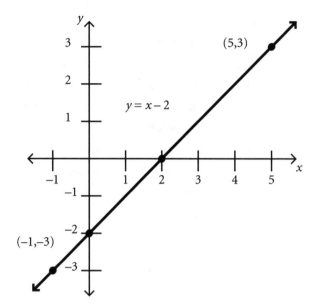

$(5, 3)$

$y = x - 2$

$(-1, -3)$

**28.**

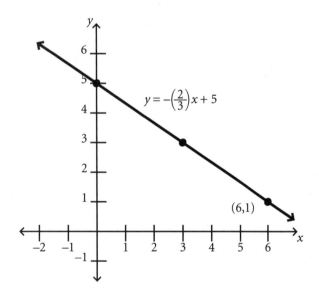

$y = -\left(\frac{2}{3}\right)x + 5$

$(6, 1)$

**30.**

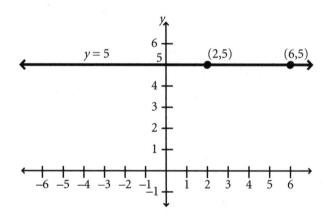

$y = 5$

$(2, 5)$

$(6, 5)$

▶ **Lesson 3**

**1.** $2^5 = 32$

**2.** $4^3 = 64$

**3.** $10^4 = 10,000$

**4.** $6^{-2} = \dfrac{1}{36}$

**5.** 1

**6.** $3^4 = 81$

**7.** 9

**8.** $\sqrt{25} = 5$

**9.** $\dfrac{1}{5}$

**10.** $\sqrt[3]{8} = 2$

**11.** $\dfrac{1}{2^3} = \dfrac{1}{8}$

**12.** Because $\dfrac{2}{3} = \dfrac{1}{3} \cdot 2$, we can calculate $8^{\frac{2}{3}} = (8^{\frac{1}{3}})^2 = (\sqrt[3]{8})^2 = 2^2 = 4$ or else $8^{\frac{2}{3}} = (8^2)^{\frac{1}{3}} = (64)^{\frac{1}{3}} = \sqrt[3]{64} = 4$.

**13.** $\dfrac{1}{\frac{1}{5}} = 5$

**14.** $\dfrac{1}{100,000}$

**15.** $8^2 = 64$

**16.** $\dfrac{1}{\frac{1}{\sqrt{16}}} = \dfrac{1}{\frac{1}{4}} = 4$

**17.** $e^{11}$

**18.** $e^7$

**19.** 1

**20.** 2

**21.** 5

**22.** $\ln(7 \cdot 2) = \ln(14)$

**23.** $\ln\left(\dfrac{24}{6}\right) = \ln(4)$

**24.** $x = \dfrac{\ln(10)}{\ln(2)}$

**25.** $x = \dfrac{\ln(11)}{\ln(8)}$

**26.** $x = \dfrac{\ln(100)}{\ln(3)} - 5$

▶ **Lesson 4**

**1.** $\dfrac{\pi}{6}$

**2.** $\pi$

**3.** $\dfrac{3\pi}{2}$

**4.** $\dfrac{5\pi}{2}$

**5.** $\dfrac{5\pi}{3}$

**6.** $60°$

**7.** $90°$

**8.** $360°$

**9.** $18°$

**10.** $144°$

**11.** 1

**12.** $\sqrt{3}$

**13.** 2

**14.** 2

**15.** $\dfrac{1}{\sqrt{3}} = \dfrac{\sqrt{3}}{3}$

**16.** $\sqrt{3}$

**17.** $\dfrac{2}{\sqrt{3}} = \dfrac{2\sqrt{3}}{3}$

**18.** $\sqrt{2}$

**19.** $\dfrac{\sqrt{3}}{3}$

**20.** $\dfrac{2\sqrt{3}}{3}$

**21.** $\dfrac{\sqrt{2}}{2}$

**22.** $-\dfrac{\sqrt{2}}{2}$

**23.** $-1$

**24.** $-\sqrt{2}$

**25.** $\sqrt{2}$

**26.** $-1$

**27.** 0

**28.** undefined

**29.** $\sqrt{3}$

**30.** $-\dfrac{\sqrt{3}}{2}$

**31.** $-\dfrac{1}{2}$

**32.** $\sqrt{3}$

**33.** $-2$

**34.** $-\dfrac{2\sqrt{3}}{3}$

**35.** $\dfrac{\sqrt{3}}{3}$

**36.** $\sqrt{2}$

### ▶ Lesson 5

**1.** 1

**2.** 1

**3.** 1

**4.** not defined

**5.** no

**6.** $-1$

**7.** 4

**8.** not defined

**9.** $-1$

**10.** yes

**11.** 1

**12.** 3

**13.** $\infty$

**14.** $\infty$

**15.** 2

**16.** $-2$

**17.** 16

**18.** $\dfrac{4}{23}$

**19.** 0

**20.** $\dfrac{\frac{1}{2}}{\frac{\pi}{6}} = \dfrac{3}{\pi}$

**21.** $2x + 1$

**22.** $3x^2$

**23.** $-\infty$

**24.** $\infty$

**25.** $\dfrac{1}{6}$

**26.** $-\infty$

**27.** $\dfrac{7}{2}$

**28.** $-\infty$

**29.** 8

**30.** $\dfrac{1}{4}$

**31.** 1

**32.** $\dfrac{1}{4}$

**33.** $-\infty$

**34.** $\dfrac{1}{260}$

**35.** $2x$

**36.** $\dfrac{1}{2\sqrt{x}}$

## ▶ Lesson 6

**1.** $f'(x) = \lim\limits_{a \to 0} \dfrac{8(x + a) + 2 - (8x + 2)}{a}$

$= \lim\limits_{a \to 0} \dfrac{8x + 8a + 2 - 8x - 2}{a} = 8$

**2.** $h'(x) = \lim\limits_{a \to 0} \dfrac{(x + a)^2 + 5 - (x^2 + 5)}{a}$

$= \lim\limits_{a \to 0} \dfrac{x^2 + 2xa + a^2 + 5 - x^2 - 5}{a}$

$= \lim\limits_{a \to 0} 2x + a = 2x$

**3.** $g'(x) = \lim\limits_{a \to 0} \dfrac{g(x + a) - g(x)}{a}$

$= \lim\limits_{a \to 0} \dfrac{10 - 10}{a} = \lim\limits_{a \to 0} \dfrac{0}{a} = 0$

**4.** $g'(x) = \lim\limits_{a \to 0} \dfrac{3 - 5(x + a) - (3 - 5x)}{a}$

$= \lim\limits_{a \to 0} \dfrac{-5a}{a} = -5$

**5.** $f'(x) = \lim\limits_{a \to 0} \dfrac{3\sqrt{x + a} - 3\sqrt{x}}{a}$

$= \lim\limits_{a \to 0} \left( \dfrac{3\sqrt{x + a} - 3\sqrt{x}}{a} \right)$

$\left( \dfrac{3\sqrt{x + a} + 3\sqrt{x}}{3\sqrt{x + a} + 3\sqrt{x}} \right)$

$= \lim\limits_{a \to 0} \dfrac{9(x + a) - 9x}{a(3\sqrt{x + a} + 3\sqrt{x})}$

$= \lim\limits_{a \to 0} \dfrac{9}{3\sqrt{x + a} + 3\sqrt{x}} = \dfrac{9}{3\sqrt{x} + 3\sqrt{x}}$

$= \dfrac{3}{2\sqrt{x}}$

**6.** $k'(x) = \lim\limits_{a\to 0} \dfrac{(x+a)^3 - x^3}{a}$

$= \lim\limits_{a\to 0} \dfrac{x^3 + 3x^2a + 3xa^2 + a^3 - x^3}{a}$

$= \lim\limits_{a\to 0} \dfrac{(3x^2 + 3xa + a^2)a}{a}$

$= \lim\limits_{a\to 0} 3x^2 + 3xa + a^2 = 3x^2$

**7.** $f'(x) = \lim\limits_{a\to 0} \dfrac{3(x+a)^2 + (x+a) - (3x^2 + x)}{a}$

$= \lim\limits_{a\to 0} \dfrac{6xa + 3a^2 + a}{a}$

$= \lim\limits_{a\to 0} 6x + 3a + 1 = 6x + 1.$

Thus, at $x = 2$, the slope is $f'(2) = 13$.

**8.** $g'(x) = \lim\limits_{a\to 0} \dfrac{(x+a)^2 - 4(x+a) +}{a}$

$= \dfrac{1 - (x^2 - 4x + 1)}{a}$

$= \lim\limits_{a\to 0} \dfrac{2xa + a^2 - 4a}{a}$

$= \lim\limits_{a\to 0} 2x + a - 4 = 2x - 4.$

Thus, there is a slope of zero when $g'(x) = 2x - 4 = 0$. This happens when $x = 2$.

**9.** $h'(x) = \lim\limits_{a\to 0} \dfrac{1 - (x+a)^2 - (1 - x^2)}{a}$

$= \lim\limits_{a\to 0} \dfrac{-2xa - a^2}{a}$

$= \lim\limits_{a\to 0} -2x - a = -2x.$

The slope at $(2,-3)$ is $h'(2) = -4$, so the equation of the tangent line is $y = -4(x-2) - 3 = -4x + 5$.

**10.** $k'(x) =$

$\lim\limits_{a\to 0} \dfrac{5(x+a)^2 + 2(x+a) - (5x^2 + 2x)}{a} =$

$\lim\limits_{a\to 0} \dfrac{10xa + 5a^2 + 2a}{a} = \lim\limits_{a\to 0} 10x + 5a + 2 =$

$10x + 2$. The $y$-value at $x = 1$ is $k(1) = 7$. The slope at $x = 1$ is $k'(1) = 12$. The equation of the tangent line is $y = 12(x-1) + 7 = 12x - 5$.

▶ **Lesson 7**

**1.** $f'(x) = 5x^4$

**2.** $\dfrac{dy}{dx} = 7x^6$

**3.** $g'(u) = -5u^{-6} = \dfrac{-5}{u^6}$

**4.** $h'(x) = 0$

**5.** $\dfrac{dy}{dt} = 4t^3$

**6.** $\dfrac{dy}{dx} = \dfrac{7}{5}\cdot x^{\frac{2}{5}}$

**7.** $f'(x) = 100x^{99}$

**8.** $f'(t) = 0$

**9.** $h'(x) = 1\cdot x^0 = 1$

**10.** $\dfrac{dy}{dx} = \dfrac{2}{3}x^{-\frac{1}{3}} = \dfrac{2}{3\sqrt[3]{x}}$

**11.** $g'(x) = -\dfrac{4}{5}x^{-\frac{9}{5}} = -\dfrac{4}{5x^{\frac{9}{5}}}$

**12.** $k(x) = x^{\frac{1}{4}}$, so $k'(x) = \dfrac{1}{4}x^{-\frac{3}{4}} = \dfrac{1}{4x^{\frac{3}{4}}}$

**13.** $y = u^{\frac{1}{2}}$, so $\dfrac{dy}{du} = \dfrac{1}{2}u^{-\frac{1}{2}} = \dfrac{1}{2\sqrt{u}}$

**14.** $y = x^{-1}$, so $\dfrac{dy}{dx} = -x^{-2} = -\dfrac{1}{x^2}$

**15.** $f(x) = \dfrac{1}{x^{\frac{1}{2}}} = x^{-\frac{1}{2}}$, so $f'(x) = -\dfrac{1}{2}x^{-\frac{3}{2}} = -\dfrac{1}{2x^{\frac{3}{2}}}$

**16.** $g(x) = x^{-3}$, so $g'(x) = -3x^{-4} = -\dfrac{3}{x^4}$

**17.** $\dfrac{dy}{dx} = 21x^6$

**18.** $f'(x) = 30x^{-11} = \dfrac{30}{x^{11}}$

**19.** $V'(r) = 4\pi r^2$

**20.** $g'(t) = \dfrac{48}{5}t^3$

**21.** $k'(x) = -2x$

**22.** $\dfrac{dy}{dt} = 12t^2 - 8$

**23.** $f'(x) = 24x^2 + 6x$

**24.** $\dfrac{dy}{dx} = 2x - 3$

**25.** $s'(t) = -32t + 5$

**26.** $F'(x) = 600x^{99} + 500x^{49} - 100x^{24} + 20x^9$

**27.** $g'(x) = \dfrac{3}{5}x^{-\frac{4}{5}} + 15x^2 = \dfrac{3}{5x^{\frac{4}{5}}} + 15x^2$

**28.** $h'(u) = 5u^4 + 16u^3 - 21u^2 - 4u + 8$

**29.** $\dfrac{dy}{dx} = -2x^{-2} - 2x^{-3} = -\dfrac{2}{x^2} - \dfrac{2}{x^3}$

**30.** $\dfrac{dy}{du} = 2u + 2u^{-3} = 2u + \dfrac{2}{u^3}$

**31.** $f'(x) = 8x - 8 - \dfrac{3}{x^2}$

**32.** $\dfrac{dy}{dx} = 2x^{-\frac{1}{2}} + 3x^{-\frac{2}{3}} = \dfrac{2}{\sqrt{x}} + \dfrac{3}{x^{\frac{2}{3}}}$

**33.** $f'(x) = -x^{-2}$, $f''(x) = 2x^{-3}$, $f'''(x) = -6x^{-4}$, and $f''''(x) = 24x^{-5}$

**34.** $s''(t) = -32$

**35.** $\dfrac{d^3y}{dx^3} = 240x - 42$

**36.** $\dfrac{dy}{dt} = 2t^{-\frac{2}{3}}$, $\dfrac{d^2y}{dt^2} = -\dfrac{4}{3}t^{-\frac{5}{3}}$, and $\dfrac{d^3y}{dt^3} = \dfrac{20}{9}t^{-\frac{8}{3}}$

## ▶ Lesson 8

**1.** pay rate in dollars per hour
**2.** fuel economy in miles per gallon
**3.** baby's growth rate in pounds per month
**4.** sunflower's growth rate in inches per week
**5.** increasing by 1 foot per year
**6.** decreasing by 6 feet per day
**7.** The profit is increasing by $3,750 per car, so the company would increase its profits if it made more cars.
**8.** $C'(3) = 4.8 - \dfrac{8}{3} \approx 2.13$, so the cost would increase by about $2.13 if the width were increased by an inch. This indicates that the cost to make the container would be cheaper if $x$ were decreased from 3.
**9.** After 3 seconds, it is at $s(3) = 75$ meters from the start. At that moment, it is traveling at $v(3) = 49$ meters per second and accelerating at $a(3) = 22$ meters per second per second.

**10.** The position function is $s(t) = -16t^2 + 64$, the velocity function is $v(t) = -32t$, and the acceleration is a constant $a(t) = -32$. It will hit the ground when $t = 2$ seconds and be traveling at $v(2) = -64$ feet per second (downward) at that instant.

**11.** The position function is $s(t) = -16t^2 + 800t$ and the velocity function is $v(t) = -32t + 800$. The bullet will stop in the air when the velocity is zero. This happens at $t = 25$ seconds, when the bullet is $s(25) = 10,000$ feet in the air.

**12.** The position function is $s(t) = -16t^2 - 10t + 1,000$, so after 4 seconds, it is $s(4) = 704$ feet off the ground. It has therefore fallen 296 feet by this moment. It is traveling $v(4) = -138$ feet per second (downward) at this moment.

**13.** $\dfrac{dy}{dx} = 20x^4 - 10\sin(x)$

**14.** $f'(t) = 3\cos(t) - \dfrac{2}{t^2}$

**15.** $g'(x) = 8 + \sin(x)$

**16.** $r'(\theta) = \dfrac{1}{2}\cos(\theta) - \dfrac{1}{2}\sin(\theta)$

**17.** $h'(x) = -\sin(x)$ because $\cos(5)$ is a constant

**18.** Because $f\left(\dfrac{\pi}{2}\right) = \sin\left(\dfrac{\pi}{2}\right) + \cos\left(\dfrac{\pi}{2}\right) = 1 + 0 = 1$, the point is $\left(\dfrac{\pi}{2},1\right)$. The slope is $f'\left(\dfrac{\pi}{2}\right) = -1$, so the equation is $y = -\left(x - \dfrac{\pi}{2}\right) + 1 = -x + \dfrac{\pi}{2} + 1$.

**19.** $f'(x) = 1 + 2x + 3x^2 + e^x$

**20.** $g'(t) = \dfrac{12}{t} + 2t$

**21.** $\dfrac{dy}{dx} = -\sin(x) - 10e^x + 8$

**22.** $h'(x) = \dfrac{1}{2\sqrt{x}} - \dfrac{8}{x}$

**23.** $k'(u) = \dfrac{15}{2}x^{\frac{3}{2}} + 5e^x$

**24.** $f'(x) = e^x + \dfrac{1}{x}$, so $f''(x) = e^x - \dfrac{1}{x^2}$

**25.** $g^{(100)}(x) = 3e^x$

**26.** $f'(10) = \dfrac{1}{10}$

## ▶ Lesson 9

**1.** $f'(x) = 2x\cos(x) - \sin(x)\cdot x^2$

**2.** $\dfrac{dy}{dt} = 24t^2e^t + e^t\cdot 8t^3 = 8t^2e^t(3 + t)$

**3.** $\dfrac{dy}{dx} = \cos^2(x) - \sin^2(x)$

**4.** $g'(x) = 6x\ln(x) + \dfrac{1}{x}(3x^2) - 20x^3 = 6x\ln(x) + 3x - 20x^3$

**5.** $h'(u) = 1\cdot e^u + e^u\cdot u - e^u = ue^u$

**6.** $k'(x) = \cos(x) + 4x^3 - (2x\sin(x) + \cos(x)x^2) = \cos(x) + 4x^3 - 2x\sin(x) - x^2\cos(x)$

**7.** $\dfrac{dy}{dx} = \dfrac{8\sin(x)}{x} + 8\ln(x)\cos(x) - \sin(x)$

**8.** $h'(t) = (\sin(t) + \cos(t) \cdot t) - (\cos(t) - \sin(t) \cdot t) = \sin(t) + t\cos(t) - \cos(t) + t\sin(t)$

**9.** $\dfrac{dy}{dx} = 15x^2 - (1 \cdot \ln(x) + \dfrac{1}{x} \cdot x)$

$\qquad = 15x^2 - \ln(x) - 1$

**10.** $f'(x) = \cos(x)\sin(x) + \cos(x)\sin(x)$

$\qquad = 2\cos(x)\sin(x)$

**11.** $\dfrac{dy}{dx} = e^x\sin(x) + (e^x\sin(x) + \cos(x)e^x) \cdot x$

**12.** $g'(x) = 12x^3\ln(x)\cos(x) +$

$\qquad \left(\dfrac{1}{x}\cos(x) - \sin(x)\ln(x)\right) \cdot 3x^4$

$\qquad = 12x^3\ln(x)\cos(x) + 3x^3\cos(x) - 3x^4\sin(x)\ln(x)$

**13.** $f'(0) = 2(0)e^0 + e^0(0)^2 + 1 = 1$, so the slope is 1.

**14.** $y = -\pi(x - \pi) = -\pi x + \pi^2$

**15.** $h'(x) = \dfrac{(3x^2 + 10)(3x^2 + 5x + 2)}{(3x^2 + 5x + 2)^2} -$

$\qquad \dfrac{(6x + 5)(x^3 + 10x + 7)}{(3x^2 + 5x + 2)^2}$

**16.** $\dfrac{dy}{dt} =$

$\dfrac{(4e^t + 1)(t^3 + 2t + 1) - (3t^2 + 2)(4e^t + t)}{(t^3 + 2t + 1)^2}$

**17.** $f'(x) = \dfrac{(1 + \frac{1}{x})(e^x - 1) - e^x(x + \ln(x))}{(e^x - 1)^2}$

**18.** $\dfrac{dy}{dx} = \dfrac{5x^4\ln(x) - \frac{1}{x} \cdot x^5}{(\ln(x))^2} = \dfrac{5x^4\ln(x) - x^4}{(\ln(x))^2}$

**19.** $f'(x) = \dfrac{2x(x^2 + 1) - 2x(x^2 - 1)}{(x^2 + 1)^2} = \dfrac{4x}{(x^2 + 1)^2}$

**20.** $g'(t) = \dfrac{15t^2\sin(t) - 5t^3\cos(t)}{25\sin^2(t)}$

**21.** $\dfrac{dy}{dx} = \dfrac{-2}{(x - 1)^2}$

**22.** $g'(u) = \dfrac{\cos(u)(u^3 - e^u) - (3u^2 - e^u)\sin(u)}{(u^3 - e^u)^2}$

**23.** $\dfrac{dy}{dx} = \dfrac{(2x + 2 + e^x)(\sin(x) + 1)}{(\sin(x) + 1)^2} -$

$\qquad \dfrac{\cos(x)(x^2 + 2x + e^x)}{(\sin(x) + 1)^2}$

**24.** $h'(t) = \dfrac{(\frac{1}{t} + 1) \cdot t^2 - 2t(\ln(t) + t)}{t^4}$

$\qquad = \dfrac{1 - t - 2\ln(t)}{t^3}$

**25.** $\dfrac{dy}{dx} = \dfrac{(\ln(x) + 1)e^x - xe^x\ln(x)}{e^x \cdot e^x}$

$\qquad = \dfrac{\ln(x) + 1 - x\ln(x)}{e^x}$

**26.** $f'(x) = \dfrac{(2xe^x + e^x \cdot x^2)\cos(x) + \sin(x) \cdot x^2e^x}{\cos^2(x)}$

**27.** $\dfrac{d^2y}{dx^2} = \dfrac{10x - 40}{(x^2 - 8x + 16)^2}$

**28.** The slope is $f'(5) = \dfrac{4}{5}$.

**29.** $\dfrac{d}{dx}(\csc(x)) = \dfrac{d}{dx}\left(\dfrac{1}{\sin(x)}\right)$

$\qquad = \dfrac{-\cos(x)}{\sin^2(x)} = -\csc(x)\cot(x)$

**30.** $\dfrac{d}{dx}(\cot(x)) = \dfrac{d}{dx}\left(\dfrac{\cos(x)}{\sin(x)}\right)$

$= \dfrac{-\sin^2(x) - \cos^2(x)}{\sin^2(x)}$

$= \dfrac{-1}{\sin^2(x)} = -\csc^2(x)$

**31.** $f'(x) = \tan(x) + x \cdot \sec^2(x)$

**32.** $g'(x) = \dfrac{\frac{1}{2\sqrt{x}}\sec(x) - \sec(x)\tan(x)\sqrt{x}}{\sec^2(x)}$

## ► Lesson 10

**1.** $f'(x) = 4(8x^3 + 7)^3 \cdot (24x^2)$

**2.** $\dfrac{dy}{dx} = 3(x^2 + 8x + 9)^2 \cdot (2x + 8)$

**3.** $h'(t) = 10(t^8 - 9t^3 + 3t + 2) \cdot$
$(8t^7 - 27t^2 + 3)$

**4.** $\dfrac{dy}{du} = \dfrac{7}{2}(u^5 - 3u^4 + 7)^{\frac{5}{2}} \cdot (5u^4 - 12u^3)$

**5.** $g'(x) = \dfrac{1}{2}(x^2 + 9x + 1)^{-\frac{1}{2}} \cdot (2x + 9)$

$= \dfrac{2x + 9}{2\sqrt{x^2 + 9x + 1}}$

**6.** $\dfrac{dy}{dx} = \dfrac{1}{3}(e^x + 1)^{-\frac{2}{3}} \cdot e^x = \dfrac{e^x}{3(e^x + 1)^{\frac{2}{3}}}$

**7.** $f'(x) = \cos(x^2) \cdot 2x$

**8.** $g'(x) = 2\sin(x) \cdot \cos(x)$

**9.** $\dfrac{dy}{dt} = \dfrac{3}{3t + 5}$

**10.** $h'(x) = -3\sin(3x)$

**11.** $\dfrac{dy}{dx} = e^{(x^2)} \cdot 2x$

**12.** $\dfrac{dy}{dx} = \dfrac{1}{x + 1}$

**13.** $s'(u) = -5\cos^4(u)\sin(u)$

**14.** $\dfrac{dy}{dx} = \dfrac{5(\ln(x))^4}{x}$

**15.** $f'(x) = e^x + 2e^{2x} + 3e^{3x}$

**16.** $\dfrac{dy}{dx} = \sec^2(e^x) \cdot e^x$

**17.** $g(x) = \dfrac{1}{2}(e^x - e^{-x})$, so $g'(x) = \dfrac{1}{2}(e^x + e^{-x})$

**18.** $f'(\theta) = \dfrac{\cos(2\theta) \cdot 2 \cdot \theta - 1 \cdot \sin(2\theta)}{\theta^2}$

$= \dfrac{2\theta\cos(\theta) - \sin(2\theta)}{\theta^2}$

**19.** $\dfrac{dy}{dx} = e^{2x} + 2xe^{2x}$

**20.** $f'(x) =$
$\sec(10x^2 + e^x)\tan(10x^2 + e^x) \cdot (20x + e^x)$

**21.** $f'(x) = 3\cos^2(8x) \cdot (-\sin(8x)) \cdot 8$

$= -24\cos^2(8x)\sin(8x)$

**22.** $\dfrac{dy}{dx} = 4(e^{9x^2 + 2x + 1})^3 \cdot (e^{9x^2 + 2x + 1}) \cdot (18x + 2)$

**23.** $g'(t) = \dfrac{\sec^2(e^t + 1) \cdot e^t}{\tan(e^t + 1)}$

**24.** $\dfrac{dy}{dx} = \cos(\sin(\sin(x))) \cdot \cos(\sin(x)) \cdot \cos(x)$

**25.** $k'(u) = \sec(\ln(8u^3))\tan(\ln(8u^3)) \cdot \dfrac{1}{8u^3} \cdot 24u^2$

$= \dfrac{3}{u}\sec(\ln(8u^3))\tan(\ln(8u^3))$

**26.** $h'(x) = \dfrac{1}{\cos(x + e^{3x})} \cdot$

$(-\sin(x + e^{3x})) \cdot (1 + e^{3x} \cdot 3)$

$= \dfrac{-(1 + 3e^{3x})\sin(x + e^{3x})}{\cos(x + e^{3x})}$

## ▶ Lesson 11

**1.** $3(y+1)^2 \cdot \dfrac{dy}{dx} = 4x^3 - 8$, so $\dfrac{dy}{dx} = \dfrac{4x^3 - 8}{3(y+1)^2}$

**2.** $3y^2 \cdot \dfrac{dy}{dx} + \dfrac{dy}{dx} = \cos(x)$, so $\dfrac{dy}{dx} = \dfrac{\cos(x)}{3y^2 + 1}$

**3.** $\dfrac{dy}{dx} = \dfrac{4}{\cos(y)} = 4\sec(y)$

**4.** $\dfrac{dy}{dx} = \dfrac{\frac{1}{x}}{1 - \frac{1}{2\sqrt{y}}} = \dfrac{2\sqrt{y}}{2x\sqrt{y} - x}$

**5.** $\dfrac{dy}{dx} = \dfrac{12x^3 - 1}{2y - 8}$

**6.** $\dfrac{dy}{dx} = \dfrac{3x^2 - e^x}{e^y}$

**7.** $\dfrac{dy}{dx} = \dfrac{-\sin(x)}{\sec^2(y)} = -\sin(x)\cos^2(y)$

**8.** $\dfrac{dy}{dx} = \dfrac{1}{2\sqrt{x+y}}\left(1 + \dfrac{dy}{dx}\right)$, so

$\dfrac{dy}{dx} = \dfrac{\frac{1}{2\sqrt{x+y}}}{1 - \frac{1}{2\sqrt{x+y}}} = \dfrac{1}{2\sqrt{x+y} - 1}$

**9.** $\dfrac{dy}{dx} = \dfrac{1 - \cos(x)}{-\cos(y)}$

**10.** $\dfrac{dy}{dx} = \dfrac{30x^2 - 12x}{1 - \frac{1}{y}} = \dfrac{30x^2 y - 12xy}{y - 1}$

**11.** $\dfrac{dy}{dx} = \dfrac{5}{2(y + x^2)^3} - 2x$

**12.** $\dfrac{dy}{dx} = \dfrac{4x^3 - 2xy}{x^2 - 4y^3}$

**13.** $\dfrac{y - x \cdot \frac{dy}{dx}}{y^2} + y + \dfrac{dy}{dx} \cdot x = 1 + \dfrac{dy}{dx}$, so

$\dfrac{dy}{dx} = \dfrac{1 - y - \frac{1}{y}}{-\frac{x}{y^2} + x - 1} = \dfrac{y^2 - y^3 - y}{-x + xy^2 - y^2}$

**14.** $\dfrac{dy}{dx} = \dfrac{3x^2\cos(y)}{\sec(y)\tan(y) + 9 + x^3\sin(y)}$

**15.** $3y^2 \cdot \dfrac{dy}{dx} + 2x = 2y \cdot \dfrac{dy}{dx} - 5 \cdot \dfrac{dy}{dx}$, so at $(-3,1)$,

the tangent slope is $\dfrac{dy}{dx} = 1$.

**16.** $\dfrac{dy}{dx} = -\dfrac{4}{9}$ at $(1,-2)$

**17.** $\dfrac{dy}{dx} = 1$ at $(4,2)$

**18.** $\dfrac{dy}{dx} = -\dfrac{21}{16}$ at $(2,3)$

**19.** $\dfrac{dy}{dx} = \dfrac{2\sqrt{3}}{3}$ at $\left(\dfrac{1}{2}, \dfrac{\pi}{6}\right)$, so the tangent equation

is $y = \dfrac{2\sqrt{3}}{3}\left(x - \dfrac{1}{2}\right) + \dfrac{\pi}{6}$.

**20.** $\dfrac{dy}{dx} = -\dfrac{8}{3}$ at $(3, -2)$, so the tangent equation is

$y = -\dfrac{8}{3}(x - 3) - 2 = -\dfrac{8}{3}x + 6$.

## ▶ Lesson 12

**1.** $\dfrac{dy}{dt} = 5(x^3 + x - 1)^4 \cdot \left(3x^2\dfrac{dx}{dt} + \dfrac{dx}{dt}\right)$

**2.** $4y^3\dfrac{dy}{dt} - 6x\dfrac{dx}{dt} = -\sin(y)\dfrac{dy}{dt}$

**3.** $3y^2\dfrac{dy}{dt} - \dfrac{dy}{dt} = 12x^3\dfrac{dx}{dt} - 20x\dfrac{dx}{dt} + 3\dfrac{dx}{dt}$

**4.** $\dfrac{1}{2\sqrt{x}} \cdot \dfrac{dx}{dt} + \dfrac{1}{2\sqrt{y}} \cdot \dfrac{dy}{dt} = 30x^2 \cdot \dfrac{dx}{dt} - 7 \cdot \dfrac{dx}{dt}$

**5.** $\dfrac{1}{y} \cdot \dfrac{dy}{dt} + e^x \cdot \dfrac{dx}{dt} = 2x \cdot \dfrac{dx}{dt} \cdot y^2 + 2y \cdot \dfrac{dy}{dt} \cdot x^2$

**6.** $10x \cdot \dfrac{dx}{dt} + 2 \cdot \dfrac{dx}{dt} = 2w \cdot \dfrac{dw}{dt}$

**7.** $\dfrac{dz}{dt} = \dfrac{4}{5}x \cdot \dfrac{dx}{dt} + \dfrac{4}{5}y \cdot \dfrac{dy}{dt} - \dfrac{3}{5x^2} \cdot \dfrac{dx}{dt}$

**8.** $2A \cdot \dfrac{dA}{dt} + 2B \cdot \dfrac{dB}{dt} = 2C \cdot \dfrac{dC}{dt}$

**9.** $\dfrac{dV}{dt} = 4\pi r^2 \cdot \dfrac{dr}{dt}$

**10.** $\dfrac{dA}{dt} = 8\pi r \cdot \dfrac{dr}{dt}$

**11.** $\dfrac{dC}{dt} = 2\pi \cdot \dfrac{dr}{dt}$

**12.** $\dfrac{dA}{dt} = \dfrac{1}{2} \cdot \dfrac{db}{dt} \cdot h + \dfrac{dh}{dt} \cdot \dfrac{1}{2}b$

**13.** $\dfrac{dx}{dt} = -\dfrac{35}{12}$

**14.** $\dfrac{dy}{dt} = -\dfrac{4}{3}$

**15.** $\dfrac{dK}{dt} = 24$

**16.** $\dfrac{dB}{dt} = 28$

**17.** Because $\dfrac{dA}{dt} = 172$, $A$ is increasing at the rate of 172 square feet per minute when $I = 20$.

**18.** $\dfrac{dR}{dt} = -\dfrac{27}{64}$, so $R$ is decreasing at the rate of $\dfrac{27}{64}$ per hour at this instant.

**19.** $\dfrac{dA}{dt} = -25$, so the area is decreasing at the rate of 25 square feet per minute.

**20.** $\dfrac{dA}{dt} = -320\pi\dfrac{\text{in}^2}{\text{min}}$, so the area shrinks by $320\pi$ square inches per minute.

**21.** $\dfrac{dr}{dt} = \dfrac{5}{2\pi}$, so the radius grows at $\dfrac{5}{2\pi} \approx 0.796$ feet per hour.

**22.** $V = s^3$, so $\dfrac{dV}{dt} = 3s^2\dfrac{ds}{dt}$, thus $\dfrac{ds}{dt} = 4$ when $\dfrac{dV}{dt} = 1{,}200$ and $s = 10$. Each side is growing at the rate of 4 inches per minute.

**23.** $\frac{db}{dt} = 7$, so the base increases at the rate of 7 inches per hour.

**24.** If the height is $y$ and the base is $x$, then $x^2 + y^2 = 10^2$ and $2x\frac{dx}{dt} + 2y\frac{dy}{dt} = 0$. After 6 seconds, $y = 6$ and $x^2 + 6^2 = 100$, so $x = 8$. Because $\frac{dy}{dt} = 1$, $2(8)\frac{dx}{dt} + 2(6)(1) = 0$, so $\frac{dx}{dt} = -\frac{3}{4}$. The end of the board is moving at the rate of $\frac{3}{4}$ of a foot each hour along the ground.

**25.** If the base is $x$ and the hypotenuse (length of the string) is $s$, then $x^2 + 100^2 = s^2$. Using this, when $s = 260$, $x$ must be 240. Because $\frac{dx}{dt} = 13$, we can calculate that $\frac{ds}{dt} = 12$. Thus, the string must be let out at 12 feet per second.

▶ **Lesson 13**

**1.** 0

**2.** $\frac{4}{5}$

**3.** $\frac{5}{2}$

**4.** $\infty$

**5.** 0

**6.** $-\frac{8}{9}$

**7.** $-\infty$

**8.** $\infty$

**9.** 1

**10.** 0

**11.** 3

**12.** $\infty$

**13.** $-\infty$

**14.** $\infty$

**15.** $\frac{4}{15}$

**16.** 0

**17.** $\infty$

**18.** 0

**19.** 0

**20.** $-\frac{4}{7}$

**21.** vertical asymptote at $x = 4$, horizontal asymptote at $y = 1$, sign diagram:

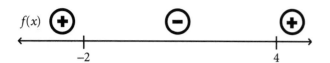

**22.** vertical asymptotes at $x = 2$ and $x = -2$, horizontal asymptote at $y = 0$, sign diagram:

**23.** vertical asymptote at $x = -3$, horizontal asymptote at $y = 1$, sign diagram:

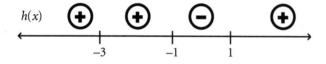

**24.** vertical asymptotes at $x = 1$ and $x = 3$, horizontal asymptote at $y = 0$, sign diagram:

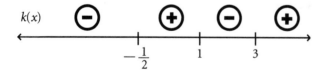

**25.** $\infty$

**26.** $\infty$

**27.** $\infty$

**28.** $-\infty$

## ▶ Lesson 14

**1.** $f(x)$ has no asymptotes. $f'(x) = 2x - 30$, thus there is a local minimum at $(15, -210)$. Because $f''(x) = 2$, the graph is always concave up.

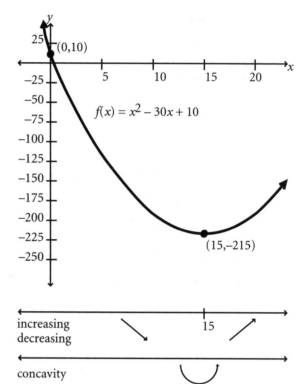

**2.** $g(x)$ has no asymptotes.

$g'(x) = -4 - 2x = -2(2 + x)$, so there is a local maximum at $(-2,4)$. Because $g''(x) = -2$, the graph is always concave down.

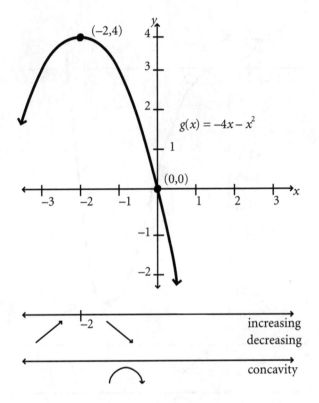

$g(x) = -4x - x^2$

increasing
decreasing

concavity

**3.** $h(x)$ has no asymptotes.

$h'(x) = 6x^2 - 6x - 36 = 6(x - 3)(x + 2)$, so there is a local maximum at $(-2,49)$ and a local minimum at $(3,-76)$. Because $h''(x) = 12x - 6 = 12\left(x - \frac{1}{2}\right)$, there is a point of inflection at $\left(\frac{1}{2}, -13\frac{1}{2}\right)$.

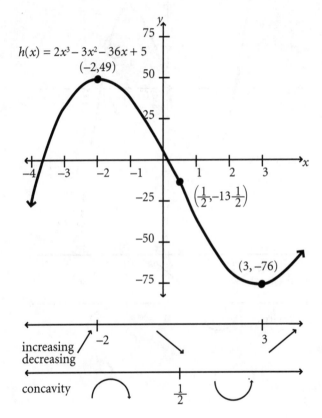

$h(x) = 2x^3 - 3x^2 - 36x + 5$

increasing
decreasing

concavity

**4.** $k(x)$ has no asymptotes.

$k'(x) = 3 - 3x^2 = 3(1-x)(1+x)$, so there is a local minimum at $(-1,-2)$ and a local maximum at $(1,2)$. $k''(x) = -6x$, so there is a point of inflection at $(0,0)$.

**5.** $f(x)$ has no asymptotes.

$f'(x) = 4x^3 - 24x^2 = 4x^2(x-6)$, so there is a local minimum at $(6,-427)$.

$f''(x) = 12x^2 - 48x = 12x(x-4)$, so there are points of inflection at $(0,-5)$ and at $(4,-251)$.

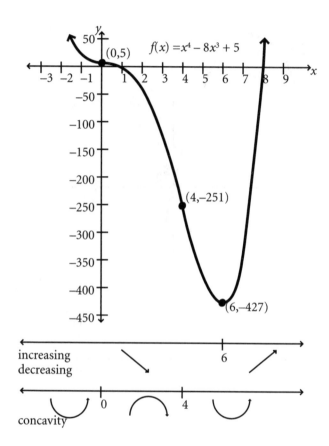

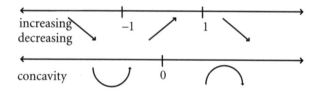

**6.** $g(x)$ has a vertical asymptote at $x = -2$ and a horizontal asymptote at $y = 1$. The first derivative is $g'(x) = \dfrac{2}{(x+2)^2}$, and the second is

$$g''(x) = \dfrac{-4}{(x+2)^3}.$$

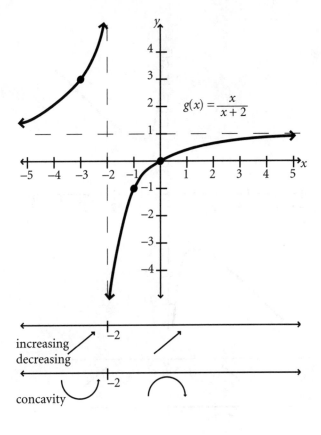

**7.** $h(x) = \dfrac{1}{(x-3)(x+3)}$ has vertical asymptotes at $x = -3$ and $x = 3$, and a horizontal asymptote at $y = 0$. Because $h'(x) = \dfrac{-2x}{(x^2-9)^2} = $

$\dfrac{-2x}{(x-3)^2(x+3)^2}$, there is a local maximum at

$\left(0, -\dfrac{1}{9}\right)$. The second derivative is

$$h''(x) = \dfrac{6x^2 + 18}{(x^2-9)^3} = \dfrac{6x^2 + 18}{(x-3)^3(x+3)^3}.$$

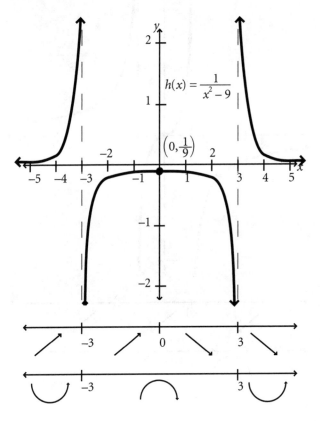

**8.** $k(x) = \dfrac{x}{(x-1)(x+1)}$ has vertical asymptotes at $x = 1$ and $x = -1$, and a horizontal asymptote at $y = 0$. The first derivative is

$k'(x) = -\dfrac{x^2 + 1}{(x^2 - 1)^2} = -\dfrac{x^2 + 1}{(x-1)^2(x+1)^2}$, and the second derivative is

$k''(x) = \dfrac{2x(x^2 + 3)}{(x-1)^3(x+1)^3}$. There is a point of inflection at $(0,0)$.

**9.** $j(x)$ has a vertical asymptote at $x = 0$ but no horizontal asymptotes. Because

$j'(x) = \dfrac{x^2 - 1}{x^2} = \dfrac{(x-1)(x+1)}{x^2}$, there is a local maximum at $(-1,-2)$ and a local minimum at $(1,2)$. The second derivative is $j''(x) = \dfrac{2}{x^3}$.

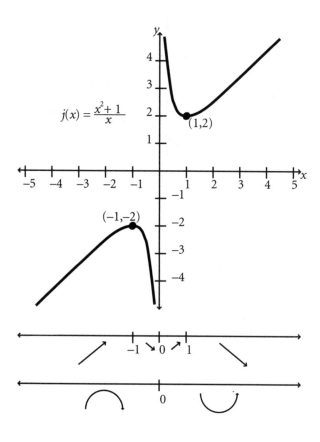

**10.** $f(x)$ has a horizontal asymptote at $y = 0$ but no vertical asymptotes. Because $f'(x) = \dfrac{1 - x^2}{(x^2 + 1)^2} = \dfrac{(1 - x)(1 + x)}{(x^2 + 1)^2}$, there is a local minimum at $\left(-1, -\dfrac{1}{2}\right)$ and a local maximum at $\left(1, \dfrac{1}{2}\right)$. Because $f''(x) = \dfrac{2x(x^2 - 3)}{(x^2 + 1)^3}$

$= \dfrac{2x(x - \sqrt{3})(x + \sqrt{3})}{(x^2 + 1)^3}$, there are points of

inflection at $\left(-\sqrt{3}, -\dfrac{\sqrt{3}}{4}\right)$, $(0,0)$, and

$\left(\sqrt{3}, \dfrac{\sqrt{3}}{4}\right)$.

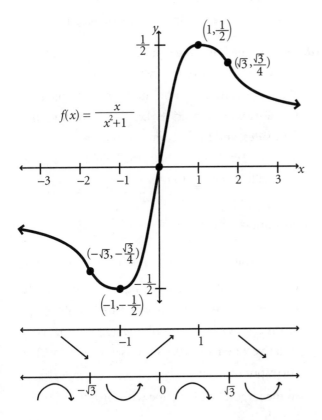

► **Lesson 15**

**1.** $P'(x) = -\dfrac{1{,}000}{x^2} + \dfrac{10{,}000}{x^3} = 0$ when $x = 10$.
Using the first derivative test, $P'(1) = 9{,}000$ is positive, so the function increases to $x = 10$ and $P'(100) = -\dfrac{1}{10} + \dfrac{1}{100} = -\dfrac{9}{100}$, so the function decreases afterward, thus $x = 10$ maximizes the profit.

**2.** $P'(x) = 18x + 40 - x^2 = -(x - 20)(x + 2)$ is zero at $x = 20$ and at $x = -2$, although making a negative number of items is impossible. Thus, $x = 20$ is the only point of slope zero. The second derivative is $P''(x) = 18 - 2x$, which is negative when $x = 20$, thus this point is an absolute maximum according to the second derivative test.

**3.** If $x$ is the number of trees beyond 30 that are planted on the acre, then the number of oranges produced will be:
Oranges$(x) = $ (number of trees) (yield per tree)
$= (30 + x)(500 - 10x) = 15{,}000 + 200x - 10x^2$.
The derivative Oranges$'(x) = 200 - 20x$ is zero when $x = 10$. Using the second derivative test, Oranges$''(x) = -20$ is negative, so this is maximal. Thus, $x = 10$ more than 30 trees should be planted, for a total of 40 trees per acre.

**4.** The total sales will be figured as follows:
Sales $= $ (number of copies)(price per copy), so
Sales$(x) = (20 + x)(100 - x)$ where $x$ is the number of copies beyond 20. The derivative Sales$'(x) = 80 - 2x$ is zero when $x = 40$. Because Sales$''(x) = -2$, this is maximal by the second derivative test. Thus, the artist should make $x = 40$ more than 20 paintings, for a total of 60 paintings in order to maximize sales.

**5.** After $x$ days, there will be $200 + 5x$ pounds of watermelon, which will valued at $90 - x$ cents per pound. Thus, the price after $x$ days will be $\text{Price}(x) = (200 + 5x)(90 - x)$ cents. The derivative is $\text{Price}'(x) = 250 - 10x$, which is zero when $x = 25$. Because $\text{Price}'(x)$ is clearly positive when $x$ is less than 25 and negative afterward, this is maximal by the first derivative test. Thus, the watermelons will fetch the highest price in 25 days.

**6.** The area is $\text{Area} = xy$ and the total fencing is $4y + 2x = 400$. Thus, $x = 200 - 2y$, so the area function can be written as follows: $\text{Area} = xy = (200 - 2y) \cdot y$. The derivative $\text{Area}'(y) = 200 - 4y$ is zero when $y = 50$. Because the second derivative is $\text{Area}''(y) = -4$, this is an absolute maximum. Thus, the optimal dimensions for the pen are $y = 50$ and $x = 200 - 2y = 200 - 2(50) = 100$.

**7.** Here, $\text{Area} = xy$ and the total fencing is $5y + x = 150$. Because $x = 150 - 5y$, the area function can be written as follows: $\text{Area}(y) = (150 - 5y)y = 150y - 5y^2$. The derivative $\text{Area}'(y) = 150 - 10y$ is zero when $y = 15$. Either the first or the second derivative test can be used to prove this is maximal. Thus, the optimal dimensions are $y = 15$ and therefore $x = 150 - 5(15) = 75$.

**8.** Suppose the height of the rectangle is $y$ and the width is $x$. The area is thus $\text{Area} = xy$, and the perimeter is $2x + 2y = 100$. Thus, $y = 50 - x$, so $\text{Area}(x) = x(50 - x) = 50x - x^2$. The derivative $\text{Area}'(x) = 50 - 2x$ is zero when $x = 25$. The second derivative is $\text{Area}''(x) = -2$, so this is maximal by the second derivative test. The height $y = 50 - x = 50 - 25 = 25$ is the same as the width. Thus, the rectangle with the largest area for a given amount of perimeter is a square.

**9.** Because $\text{Volume} = \pi r^2 h = 16\pi$, it follows that $h = \dfrac{16}{r^2}$. Thus, the surface area function is

$$\text{Area}(r) = 2\pi r^2 + 2\pi r\left(\frac{16}{r^2}\right) = 2\pi r^2 + \frac{32\pi}{r}.$$

The derivative $\text{Area}'(r) = 4\pi r - \dfrac{32\pi}{r^2}$ is zero when $4\pi r = \dfrac{32\pi}{r^2}$, so $r^3 = 8$. Thus, the only point of slope zero is when $r = 2$. The second derivative is $\text{Area}''(r) = 4\pi + \dfrac{64\pi}{r^3}$, which is positive when $r = 2$. Thus, by the second derivative test, $r = 2$ is the absolute minimum. Thus, a radius of $r = 2$ and a height of $h = \dfrac{16}{r^2} = 4$ will minimize the surface area.

**10.** Because the box has a square bottom, its length and width can be both $x$, while its height is $y$. Thus, the volume is $\text{Volume} = x^2 y$ and the surface area is $\text{Area} = x^2 + 4xy + x^2$ (the top, the four sides, and the bottom). And because $\text{Area} = 2x^2 + 4xy = 600$, the height $y = \dfrac{600 - 2x^2}{4x} = \dfrac{150}{x} - \dfrac{x}{2}$. Thus, the

$$\text{Volume} = x^2 y = x^2\left(\frac{150}{x} - \frac{x}{2}\right) = 150x - \frac{1}{2}x^3.$$

The derivative $\text{Volume}'(x) = 150 - \dfrac{3}{2}x^2$ is zero when $x^2 = 100$. Negative lengths are impossible, therefore this is zero only when $x = 10$. By the second derivative test, $\text{Volume}''(x) = -3x$ is negative when $x = 10$, so this is a maximum. The corresponding height is $y = \dfrac{150}{10} - \dfrac{10}{2} = 10$, so the largest box is a cube with all sides of length 10.

**11.** Because the box has a square bottom, let $x$ be both the length and width, and $y$ be the height. The area of each side is thus $xy$, so the cost to build four of them at ten cents a square foot is $0.10(4xy) = 0.4xy$ dollars. The area of the top is $x \cdot x = x^2$, so it will cost $x^2$ dollars to build. Similarly, it will cost $7x^2$ dollars to build the base. The total cost of the box is therefore Cost $= 0.4xy + 8x^2$. Because the volume is $x^2y = 40,000$, we know $y = \dfrac{40,000}{x^2}$. Thus, the cost function can be written:

$$\text{Cost}(x) = 0.4x\left(\frac{40,000}{x^2}\right) + 8x^2$$
$$= \frac{16,000}{x} + 8x^2.$$

The derivative:
$$\text{Cost}'(x) = -\frac{16,000}{x^2} + 16x$$
is zero only when $x^3 = 1,000$ or $x = 10$.

By the second derivative test,
$$\text{Cost}''(x) = \frac{32,000}{x^3} + 16 \text{ is positive when}$$
$x = 10$, so this is the absolute minimum. The cheapest box will be built when $x = 10$ and $y = 400$.

**12.** Inside the margins, the area is:

$$\text{Area} = \left(x - 2\left(1\frac{1}{2}\right)\right)(y - 2)$$
$$= (x - 3)(y - 2) = xy - 2x - 3y + 6.$$

The total area of the page is $xy = 96$, so $y = \dfrac{96}{x}$. Therefore,

$$\text{Area}(x) = x\left(\frac{96}{x}\right) - 2x - 3y + 6$$
$$= 102 - 2x - \frac{288}{x}.$$

The derivative $\text{Area}'(x) = -2 + \dfrac{288}{x^2}$ is zero when $x^2 = 144$, thus when $x = 12$ (ignore negative lengths). The second derivative $\text{Area}''(x) = -\dfrac{576}{x^3}$ is negative when $x = 12$,

so this is the absolute maximum. Thus, the dimensions that maximize the printed area are $x = 12$ and $y = 8$.

▶ **Lesson 16**

**1.** 2

**2.** $\dfrac{3}{2}$

**3.** $\dfrac{7}{2}$

**4.** $2\pi$

**5.** 3

**6.** $3 + 2\pi$

**7.** $-\dfrac{3}{2}$

**8.** 0

**9.** $-\dfrac{3}{2}$

**10.** 4

**11.** 0

**12.** $-\dfrac{1}{2}$

**13.** 16

**14.** 6

**15.** 0

**16.** $\dfrac{21}{2}$

**17.** 35

**18.** 48

**19.** 5

**20.** $-3$

**21.** 7

**22.** 7

**23.** $-11$

**24.** $-1$

**25.** 8

**26.** $-17$

**27.** 25

▶ **Lesson 17**

**1.** $\dfrac{5}{4}$

**2.** 3

**3.** $\dfrac{21}{4}$

**4.** 8

**5.** $\dfrac{45}{4}$

**6.** 0

**7.** 0

**8.** 7

**9.** 14

**10.** 21

**11.** 28

**12.** 35

**13.** 10

**14.** $-4$

**15.** 36

**16.** 20

**17.** $\dfrac{2}{3}$

**18.** $\dfrac{16}{3}$

**19.** $\dfrac{38}{3}$

**20.** $\dfrac{2,000}{3}$

**21.** $\dfrac{1}{2}$

**22.** $\dfrac{4}{5}$

**23.** $\dfrac{1}{4}$

**24.** $\dfrac{2}{3}$

▶ **Lesson 18**

**1.** $\dfrac{1}{6}x^6 + c$

**2.** $\dfrac{1}{13}x^{13} + c$

**3.** $\dfrac{1}{7}u^7 + c$

**4.** 72

**5.** 40

**6.** $\dfrac{1}{-2}t^{-2} + c = -\dfrac{1}{2t^2} + c$

**7.** $\dfrac{21}{2}$

**8.** $\frac{3}{8}x^{\frac{8}{3}} + c$

**9.** $\frac{4}{5}x^{\frac{5}{4}} + c$

**10.** $\frac{3}{4}u^{\frac{4}{3}} + c$

**11.** $5x + c$

**12.** $5t + c$

**13.** $40$

**14.** $\frac{1}{4}$

**15.** $\frac{9}{5}x^5 + c$

**16.** $\frac{8}{3}u^3 + c$

**17.** $\frac{1}{2}x^2 - \frac{2}{3}x^{\frac{3}{2}} + c$

**18.** $2x^3 - 5x^2 + 5x + c$

**19.** $48$

**20.** $11$

**21.** $115$

**22.** $\frac{1}{4}t^{12} + 3t^3 + \frac{1}{2}t^2 + c$

**23.** $-6$

**24.** $2x^4 + \frac{10}{3}x^3 - 2x^2 + 2x + c$

**25.** $58$

**26.** $208$

**27.** $-\frac{2}{x^2} + c$

**28.** $\frac{9}{13}x^{\frac{13}{3}} - \frac{56}{11}x^{\frac{11}{7}} + c$

**29.** $\frac{1}{3}x^3 - 5\sin(x) + c$

**30.** $3e^x + \frac{1}{2}x^4 + c$

**31.** $\ln|u| + c$

**32.** $\frac{1}{2}\theta^2 - 2\cos(\theta) + c$

**33.** $-\cos(x) + 2e^x + c$

**34.** $\left(\frac{1}{2} + e^1\right) - (0 + e^0) = e - \frac{1}{2}$

**35.** $4\ln(e) - 4\ln(1) = 4 - 0 = 4$

**36.** $4\sqrt{2}$

▶ **Lesson 19**

**1.** $\frac{1}{40}(x^5 + 1)^8 + c$

**2.** $\frac{1}{44}(4x + 3)^{11} + c$

**3.** $\frac{1}{15}$

**4.** $\frac{1}{4}x^4 - \frac{9}{2}x^2 + 4x + c$

**5.** $\frac{1}{3}(x^2 - 1)^{\frac{3}{2}} + c$

**6.** $14$

**7.** $26$

**8.** $\ln 3x^3 - 5x + c$

**9.** $\frac{1}{2}\sin(x^4) + c$

**10.** $\sqrt{3x^4 - 2x + 1} + c$

**11.** $\frac{1}{4}(4x^2 + 5x - 1)^4 + c$

**12.** $-\dfrac{1}{16(4x^2 + 5)^2} + c$

**13.** $\frac{1}{4}\ln|4x + 10| + c$

**14.** Using $u = \sin(x)$, the solution is $\frac{1}{2}\sin^2(x) + c$. Using $u = \cos(x)$, the solution is $-\frac{1}{2}\cos^2(x) + c$. Because $\sin^2(x) + \cos^2(x) = 1$, these solutions will be the same if the second $+c$ is $\frac{1}{2}$ greater than the first one.

**15.** $\frac{1}{3}\sin^3(x) + c$

**16.** $\frac{1}{4}\sin(4x) + c$

**17.** $4\sin(x) + c$

**18.** $-\frac{1}{7}\cos(7x - 2) + c$

**19.** $-\cos(e^x) + c$

**20.** $\frac{1}{4}(\ln(x))^4 + c$

**21.** $\ln|\ln(x)| + c$

**22.** $\frac{1}{2}e^{(x^2)} + c$

**23.** $-\ln(\cos(x)) + c$

**24.** $\ln(1 + e^x) + c$

▶ **Lesson 20**

**1.** $\frac{1}{6}x^6\ln(x) - \frac{1}{36}x^6 + c$, done by parts with $u = \ln(x)$

**2.** $-x\cos(x) + \sin(x) + c$, by parts with $u = x$

**3.** $-\frac{1}{2}\cos(x^2) + c$, by the substitution $u = x^2$

**4.** $(x + 3)\sin(x) + \cos(x) + c$, by parts with $u = x + 3$

**5.** $\frac{1}{2}(\ln(x))^2 + c$, by substituting $u = \ln(x)$

**6.** $-x^2\cos(x) + 2x\sin(x) + 2\cos(x) + c$, using parts twice

**7.** $\frac{1}{3}x^3 - \cos(x) + c$, by basic integration

**8.** $\frac{1}{3}e^{x^3 + 1} + c$, by substituting $u = x^3 + 1$

**9.** $x^2e^x - 2xe^x + 2e^x + c$, using parts twice

**10.** $\left(\frac{1}{4}x^4 + \frac{3}{2}x^2 - x\right)\ln(x) - \frac{1}{16}x^4 - \frac{3}{4}x^2 + x + c$, by parts with $u = \ln(x)$

**11.** $\frac{1}{2}x^2 + x\ln(x) - x + c$, evaluating $\displaystyle\int \ln(x)\,dx$ by parts

**12.** $\frac{2}{3}(x - 1)^{\frac{3}{2}} + c$, substituting $u = x - 1$

**13.** $\frac{2}{3}x(x - 1)^{\frac{3}{2}} - \frac{4}{15}(x - 1)^{\frac{5}{2}} + c$, by parts with $u = x$

**14.** $-xe^{-x} - e^{-x} + c$, by parts with $u = x$

**15.** $\frac{2}{3}x^{\frac{3}{2}}\ln(x) - \frac{4}{9}x^{\frac{3}{2}} + c$, by parts with $u = \ln(x)$

**16.** $-\dfrac{\ln(x)}{2x^2} - \dfrac{1}{4x^2} + c$, by parts with $u = \ln(x)$

**17.** $\ln|x| + c$, by basic integration

**18.** $(x^2 - 1)\sin(x) + 2x\cos(x) - 2\sin(x) + c$, using parts twice

**19.** $-e^{\frac{1}{x}} + c$, by substituting $u = \dfrac{1}{x}$

**20.** $-\dfrac{2}{3}(\cos(x))^{\frac{3}{2}} + c$, by substituting $u = \cos(x)$

**21.** $-\cos(x) \cdot \ln(\cos(x)) + \cos(x) + c$, by parts with $u = \ln(\cos(x))$. This could also be solved by substitution with $u = \cos(x)$, though it would require knowing $\displaystyle\int \ln(u)\, du$.

**22.** $\dfrac{1}{2}\big(e^x\sin(x) + e^x\cos(x)\big) + c$, by parts twice, plus the trick from the previous example

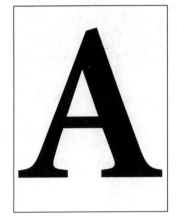

# Glossary

**acceleration** the rate at which the speed of a moving object is increasing or decreasing

**additive rule** parts of a function added together can be differentiated separately:
$$\frac{d}{dx}(f(x) + g(x)) = f'(x) + g'(x)$$

**antiderivative** given a function $f(x)$, the antiderivative is a function $g(x)$ such that $g'(x) = f(x)$.

**asymptote** a place where a graph flattens out like an infinite straight line

**Chain Rule** $\frac{d}{dx}(f(g(x))) = f'(g(x)) \cdot g'(x)$

**closed interval** the set of all the numbers between and including two endpoints, like all $x$ such that $a \le x \le b$

**composition** the process of plugging one function into another. The composition of functions $f$ and $g$ is $f \circ g(x) = f(g(x))$.

**concave down** when the graph of a function curves downward, like a frown

**concave up** when the graph of a function curves upward, like a smile

**concavity** the way a graph curves either upward or downward

**Constant Coefficient Rule** a constant $c$ multiplied in front of a function is unaffected by differentiation: $\frac{d}{dx}(c \cdot f(x)) = c \cdot f'(x)$

**Constant Rule** the derivative of a constant is zero.

**continuous** a graph is continuous between breaks.

**cosecant** abbreviated csc; see *trigonometry*

**cosine** abbreviated cos; see *trigonometry*

**cotangent** abbreviated cot; see *trigonometry*

**critical points** points of slope zero, points where the derivative is undefined, and endpoints of the domain

**decreasing** when the graph of function goes down from left to right

**definite integral** the area between a graph $y = f(x)$ and the $x$-axis from $x = a$ to $x = b$ where area below the $x$-axis counts as negative, written $\int_{b}^{a} f(x)dx$

**degrees**   measure the size of angles in such a way that a complete circle is 360°

**derivative**   the derivative of function $y = f(x)$ is
$$\frac{dy}{dx} = f'(x) = \lim_{a \to 0} \frac{f(x + a) - f(x)}{a},$$
which is the slope of the tangent line at point $(x, f(x))$.

**discontinuity**   a break in a graph

**domain**   the set of all the numbers that can be put into a function

**e**   a number approximately 2.71828 with the property that $\frac{d}{dx}(e^x) = e^x$

**explicit**   a function is explicit if its formula is known exactly.

**exponent**   an exponent says how many times a factor is multiplied by itself.

**first derivative test**   if a function increases to a point and then decreases afterward, then that point is the maximum. If the function decreases to a point and then increases afterward, then the point is the minimum.

**function**   a mathematical object that assigns one number of its range for every number in its domain

**Fundamental Theorem of Calculus**   if
$$g(x) = \int_0^x f(t)dt, \text{ then } g'(x) = f(x). \text{ Thus,}$$
$$\int_a^b f(x)dx = g(b) - g(a) \text{ where } g'(x) = f(x).$$

**graph**   a visual depiction of a function where the height of each point is the value assigned to the number on the horizontal axis

**horizontal asymptote**   where a graph flattens out to run straight off to the right or left

**implicit**   a function is implicit if it was defined in an indirect manner so that its exact formula is unknown.

**implicit differentiation**   the process of taking a derivative of both sides of an equation and using the Chain Rule with $\frac{d}{dx}(x) = 1$, $\frac{d}{dx}(y) = \frac{dy}{dx}$, $\frac{d}{dt}(x) = \frac{dx}{dt}$, and so on

**increasing**   when the graph of a function goes up from left to right

**indefinite integral**   represents the antiderivative:
$$\int f(x)dx = g(x) + c \text{ if and only if } g'(x) = f(x)$$

**integral**   see either definite integral or indefinite integral

**L'Hôpital's Rule**   If $\lim_{x \to \infty} f(x) = \pm\infty$ and $\lim_{x \to \infty} g(x) = \pm\infty$, then $\lim_{x \to \infty} \frac{f(x)}{g(x)} = \lim_{x \to \infty} \frac{f'(x)}{g'(x)}$. The same is true when $\lim_{x \to -\infty}$.

**limit**   the limit $\lim_{x \to d} f(x) = L$ means that the values of $f(x)$ get very close to $L$ as $x$ gets close to $a$.

**limit from the left**   $\lim_{x \to a^-} f(x) = L$ means that the values of $f(x)$ are close to $L$ when $x$ is close to, and less than, $a$.

**limit from the right**   $\lim_{x \to a^+} f(x) = L$ means that the values of $f(x)$ are close to $L$ when $x$ is close to, and greater than, $a$.

**limits at infinity**   $\lim_{x \to \infty} f(x) = L$ means that the values of $y = f(x)$ get close to $y = L$ as $x$ gets really big. If large negative values of $x$ are used and $y = f(x)$ gets close to $y = L$, then $\lim_{x \to -\infty} f(x) = L$.

**limits of integration**   the limits of the integral $\int_a^b f(x)dx$ are $a$ and $b$.

**local maximum**   the lowest point on a graph in that immediate area

**local minimum** the highest point on a graph in that immediate area, like a hilltop

**natural logarithm** the inverse $\ln(x)$ of the exponential function $e^x$. Thus, $y = \ln(x)$ if and only if $e^y = x$.

**oscillate** to repeatedly go back and forth across a range of values

**point of inflection** a point on a graph where the concavity changes

**point-slope formula** the equation of a straight line through $(x_1, y_1)$ with slope $m$ is
$$y = m(x - x_1) + y_1.$$

**polynomial** the sum of powers of a variable, complete with constant coefficients. For example, $x^2 + 3x - 5$ and $10x^7 - 12x^5 + 4x^2 - x$ are both polynomials.

**position function** gives the mark on a line where a moving object is at a given time

**Power Rule** $\dfrac{d}{dx}(x^n) = n \cdot x^{n-1}$

**Product Rule**
$$\frac{d}{dx}(f(x) \cdot g(x)) = f'(x) \cdot g(x) + g'(x) \cdot f(x)$$

**Pythagorean theorem** the squares of the legs of a right triangle add up to the square of the hypotenuse.

**Quotient Rule**
$$\frac{d}{dx}\left(\frac{f(x)}{g(x)}\right) = \frac{f'(x) \cdot g(x) - g'(x) \cdot f(x)}{(g(x))^2}$$

**radians** measure the size of angles in such a way that a complete circle is $2\pi$ radians

**range** the set of all the numbers that can be the value of a function

**rate of change** how fast a quantity is increasing or decreasing

**secant** abbreviated sec; see *trigonometry*

**rational function** a rational function is the quotient of two polynomials. For example, $\dfrac{8x^3 - 10x + 4}{5x - 2}$ is a rational function.

**second derivative** the derivative of the derivative

**second derivative test** a point of slope zero is the maximum if the second derivative is always negative, and a minimum if the second derivative is always positive.

**sign diagram** tells where a function is positive and negative

**sine** abbreviated sin; see *trigonometry*

**slope** the amount a straight line goes up or down with each step to the right

**slope-intercept formula** the equation of a straight line with slope $m$ that crosses the $y$-axis at $y = b$ is
$$y = mx + b.$$

**Squeeze Theorem** if $f(x) \le g(x) \le h(x)$ and $\lim\limits_{x \to d} f(x) = L = \lim\limits_{x \to a} h(x)$, then $\lim\limits_{x \to a} g(x) = L$.

**substitution** an integration technique used to reverse the Chain Rule

**tangent** abbreviated tan; see *trigonometry*

**tangent line** a straight line that indicates the direction of a curve at a given point

**third derivative** the derivative of the second derivative

**trigonometric identities** $\tan(x) = \dfrac{\sin(x)}{\cos(x)}$, $\sec(x) = \dfrac{1}{\cos(x)}$, $\csc(x) = \dfrac{1}{\sin(x)}$, $\cot(x) = \dfrac{\cos(x)}{\sin(x)}$, and $\sin^2(x) + \cos^2(x) = 1$

**trigonometry** the study of functions formed by dividing one side of a right triangle by another. When the right triangle has angle $x$, the hypotenuse has length $H$, the side adjacent to $x$ has length $A$, and the side opposite $x$ has length $O$, then

$$\sin(x) = \frac{O}{H}$$

$$\cos(x) = \frac{A}{H}$$

$$\sec(x) = \frac{H}{A}$$

$$\csc(x) = \frac{H}{O}$$

$$\tan(x) = \frac{O}{A}$$

$$\cot(x) = \frac{A}{O}$$

The derivatives are:

$$\frac{d}{dx}(\sin(x)) = \cos(x)$$

$$\frac{d}{dx}(\cos(x)) = -\sin(x)$$

$$\frac{d}{dx}(\sec(x)) = \sec(x)\tan(x)$$

$$\frac{d}{dx}(\csc(x)) = -\csc(x)\cot(x)$$

$$\frac{d}{dx}(\tan(x)) = \sec^2(x)$$

$$\frac{d}{dx}(\cot(x)) = -\csc^2(x)$$

**unit circle** the circle of radius 1 centered at the origin

**velocity** the speed of a moving object at a particular time

**vertical asymptote** where a graph looks like a straight up-and-down line

# B ▶ How to Prepare for a Test

A standardized test is nothing to fear. Many people clutch and worry about a testing situation, but you're much better off taking that nervous energy and turning it into something positive that will help you do well on your test rather than inhibit your testing ability. The following pages include valuable tips for combating test anxiety, that sinking or blank feeling some people get as they begin a test or encounter a difficult question. Next, you will find valuable tips for using your time wisely and for avoiding errors in a testing situation. Finally, you will find a plan for preparing for the test, a plan for the test day, and a suggestion for an after-test activity.

## ▶ Combating Test Anxiety

Knowing what to expect and being prepared for it is the best defense against test anxiety, that worrisome feeling that keeps you from doing your best. Practice and preparation keeps you from succumbing to that feeling. Nevertheless, even the brightest, most well-prepared test takers may suffer from occasional bouts of test anxiety. But don't worry; you can overcome it.

## Take the Test One Question at a Time

Focus all of your attention on the one question you're answering. Block out any thoughts about questions you've already read or concerns about what's coming next. Concentrate your thinking where it will do the most good—on the question you're answering.

## Develop a Positive Attitude

Keep reminding yourself that you're prepared. You've studied hard, so you're probably better prepared than most others who are taking the test. Remember, it's only a test, and you're going to do your best. That's all anyone can ask of you. If that nagging drill sergeant inside your head starts sending negative messages, combat him or her with positive ones of your own.

- "I'm doing just fine."
- "I've prepared for this test."
- "I know exactly what to do."
- "I know I can get the score I'm shooting for."

You get the idea. Remember to drown out negative messages with positive ones of your own.

## If You Lose Your Concentration

Don't worry about it! It's normal. During a long test, it happens to everyone. When your mind is stressed or overexerted, it takes a break whether you want it to or not. It's easy to get your concentration back if you simply acknowledge the fact that you've lost it and take a quick break. Your brain needs very little time (seconds really) to rest.

Put your pencil down and close your eyes. Take a few deep breaths and listen to the sound of your breathing. The ten seconds or so that this takes is really all the time your brain needs to relax and get ready to focus again.

Try this technique several times in the days before the test when you feel stressed. The more you

practice, the better it will work for you on the day of the test.

## If You Freeze before or during the Test

Don't worry about a question that stumps you even though you're sure you know the answer. Mark it and go on to the next question. You can come back to the stumper later. Try to put it out of your mind completely until you come back to it. Just let your subconscious chew on the question while your conscious mind focuses on the other items (one at a time, of course). Chances are, the memory block will be gone by the time you return to the question.

If you freeze before you begin the test, here's what to do:

1. Take a little time to look over the test.
2. Read a few of the questions.
3. Decide which ones are the easiest and start there.
4. Before long, you'll be "in the groove."

# ▶ Time Strategies

Use your time wisely to avoid making careless errors.

## Pace Yourself

The most important time strategy is to pace yourself. Before you begin, take just a few seconds to survey the test, making note of the number of questions and of the sections that look easier than the rest. Rough out a time schedule based upon the amount of time available to you. Mark the halfway point on your test and make a note beside that mark of what the time will be when the testing period is half over.

## Keep Moving

Once you begin the test, keep moving. If you work slowly in an attempt to make fewer mistakes, your mind will become bored and begin to wander. You'll end up making far more mistakes if you're not concentrating.

As long as we're talking about mistakes, don't stop for difficult questions. Skip them and move on. You can come back to them later if you have time. A question that takes you five seconds to answer counts as much as one that takes you several minutes, so pick up the easy points first. Besides, answering the easier questions first helps to build your confidence and gets you in the testing groove. Who knows? As you go through the test, you may even stumble across some relevant information to help you answer those tough questions.

## Don't Rush

Keep moving, but don't rush. Think of your mind as a seesaw. On one side is your emotional energy. On the other side is your intellectual energy. When your emotional energy is high, your intellectual capacity is low. Remember how difficult it is to reason with someone when you're angry? On the other hand, when your intellectual energy is high, your emotional energy is low. Rushing raises your emotional energy. Remember the last time you were late for work? All that rushing around caused you to forget important things—like your lunch. Move quickly to keep your mind from wandering, but don't rush and get flustered.

## Check Yourself

Check yourself at the halfway mark. If you're a little ahead, you know you're on track and may even have a little time left to check your work. If you're a little behind, you have several choices. You can pick up the pace a little, but do this only if you can do it comfortably. Remember—don't rush! You can also skip around in the remaining portion of the test to pick up as many easy points as possible. This strategy has one drawback, however. If you are marking a bubble-style

answer sheet, and you put the right answers in the wrong bubbles—they're wrong. So pay close attention to the question numbers if you decide to do this.

## ▶ Avoiding Errors

When you take the test, you want to make as few errors as possible in the questions you answer. Here are a few tactics to keep in mind.

## Control Yourself

Remember the comparison between your mind and a seesaw that you read a few paragraphs ago? Keeping your emotional energy low and your intellectual energy high is the best way to avoid mistakes. If you feel stressed or worried, stop for a few seconds. Acknowledge the feeling (Hmmm! I'm feeling a little pressure here!), take a few deep breaths, and send yourself a few positive messages. This relieves your emotional anxiety and boosts your intellectual capacity.

## Directions

In many standardized testing situations, a proctor reads the instructions aloud. Make certain you understand what is expected. If you don't, ask. Listen carefully for instructions about how to answer the questions and make certain you know how much time you have to complete the task. Write the time on your test if you don't already know how long you have to take the test. If you miss this vital information, ask for it. You need it to do well on your test.

## Answers

Place your answers in the right blanks or the corresponding ovals on the answer sheet. Right answers in the wrong place earn no points. It's a good idea to check every five to ten questions to make sure you're in the right spot. That way you won't need much time to correct your answer sheet if you have made an error.

## ▶ Reading Long Passages

Frequently, standardized tests are designed to test your reading comprehension. The reading sections often contain passages of a paragraph or more. Here are a few tactics for approaching these sections.

This may seem strange, but some questions can be answered without ever reading the passage. If the passage is short, a paragraph around four sentences or so, read the questions first. You may be able to answer them by using your common sense. You can check your answers later after you've actually read the passage. Even if you can't answer any of the questions, you know what to look for in the passage. This focuses your reading and makes it easier for you to retain important information. Most questions will deal with isolated details in the passage. If you know what to look for ahead of time, it's easier to find the information.

If a reading passage is long and is followed by more than ten questions, you may end up spending too much time reading the questions first. Even so, take a few seconds to skim the questions and read a few of the shorter ones. As you read, mark up the passage. If you find a sentence that seems to state the main idea of the passage, underline it. As you read through the rest of the passage, number the main points that support the main idea. Several questions will deal with this information. If it's underlined and numbered, you can locate it easily. Other questions will ask for specific details. Circle information that tells who, what, when, or where. The circles will be easy to locate later if you run across a question that asks for specific information. Marking up a passage in this way also heightens your concentration and makes it more likely that you'll remember the information when you answer the questions following the passage.

## Choosing the Right Answers

Make sure you understand what the question is asking. If you're not sure of what's being asked, you'll never know whether you've chosen the right answer. So figure out what the question is asking. If the answer isn't readily apparent, look for clues in the answer choices. Notice the similarities and differences in the answer choices. Sometimes, this helps to put the question in a new perspective and makes it easier to answer. If you're still not sure of the answer, use the process of elimination. First, eliminate any answer choices that are obviously wrong. Then reason your way through the remaining choices. You may be able to use relevant information from other parts of the test. If you can't eliminate any of the answer choices, you might be better off to skip the question and come back to it later. If you can't eliminate any answer choices to improve your odds when you come back later, then make a guess and move on.

## If You're Penalized for Wrong Answers

You must know whether there's a penalty for wrong answers before you begin the test. If you don't, ask the proctor before the test begins. Whether you make a guess or not depends upon the penalty. Some standardized tests are scored in such a way that every wrong answer reduces your score by one-fourth or one-half of a point. Whatever the penalty, if you can eliminate enough choices to make the odds of answering the question better than the penalty for getting it wrong, make a guess.

Let's imagine you are taking a test in which each answer has four choices and you are penalized one-fourth of a point for each wrong answer. If you have no clue and cannot eliminate any of the answer choices, you're better off leaving the question blank because the odds of answering correctly are one in four. This makes the penalty and the odds equal. However, if you can eliminate one of the choices, the odds are now in your favor. You have a one in three chance of answering the question correctly. Fortunately, few tests are scored using such elaborate means, but if your test is one of them, know the penalties and calculate your odds before you take a guess on a question.

## If You Finish Early

Use any time you have left at the end of the test or test section to check your work. First, make certain you've put the answers in the right places. As you're doing this, make sure you've answered each question only once. Most standardized tests are scored in such a way that questions with more than one answer are marked wrong. If you've erased an answer, make sure you've done a good job. Check for stray marks on your answer sheet that could distort your score.

After you've checked for these obvious errors, take a second look at the more difficult questions. You've probably heard the folk wisdom about never changing an answer. If you have a good reason for thinking a response is wrong, change it.

## ▶ The Days before the Test

To do your very best on an exam, you have to take control of your physical and mental state. Exercise, proper diet, and rest will ensure that your body works with, rather than against, your mind on exam day, as well as during your preparation.

## Physical Activity

Get some exercise in the days preceding the test. You'll send some extra oxygen to your brain and allow your thinking performance to peak on the day you take the test. Moderation is the key here. You don't want to exercise so much that you feel exhausted, but a little physical activity will invigorate your body and brain.

## Balanced Diet

Like your body, your brain needs the proper nutrients to function well. Eat plenty of fruits and vegetables in the days before the test. Foods that are high in lecithin, such as fish and beans, are especially good choices. Lecithin is a mineral your brain needs for peak performance. You may even consider a visit to your local pharmacy to buy a bottle of lecithin tablets several weeks before your test.

## Rest

Get plenty of sleep the nights before you take the test. Don't overdo it though or you'll make yourself as groggy as if you were overtired. Go to bed at a reasonable time, early enough to get the number of hours you need to function effectively. You'll feel relaxed and rested if you've gotten plenty of sleep in the days before you take the test.

## Trial Run

At some point before you take the test, make a trial run to the testing center to see how long it takes. Rushing raises your emotional energy and lowers your intellectual capacity, so you want to allow plenty of time on the test day to get to the testing center. Arriving 10 or 15 minutes early gives you time to relax and get situated.

## Test Day

It's finally here, the day of the big test. Set your alarm early enough to allow plenty of time. Eat a good breakfast. Avoid anything that's really high in sugar, such as donuts. A sugar high turns into a sugar low after an hour or so. Cereal and toast, or anything with complex carbohydrates is a good choice. Eat only moderate amounts. You don't want to take a test feeling stuffed!

Pack a high-energy snack to take with you. You may have a break sometime during the test when you can grab a quick snack. Bananas are great. They have a moderate amount of sugar and plenty of brain nutrients, such as potassium. Most proctors won't allow you to eat a snack while you're testing, but a peppermint shouldn't pose a problem. Peppermints are like smelling salts for your brain. If you lose your concentration or suffer from a momentary mental block, a peppermint can get you back on track. Don't forget the earlier advice about relaxing and taking a few deep breaths.

Leave early enough so you have plenty of time to get to the test center. Allow a few minutes for unexpected traffic. When you arrive, locate the restroom and use it. Few things interfere with concentration as much as a full bladder. Then find your seat and make sure it's comfortable. If it isn't, tell the proctor and ask to change to something you find more suitable.

Now relax and think positively! Before you know it, the test will be over, and you'll walk away knowing you've done as well as you can.

## After the Test

Two things are important for after the test:

1. Plan a little celebration.
2. Go to it.

If you have something to look forward to after the test is over, you may find it easier to prepare well for the test and to keep moving during the test. Good luck!